HEAVY QUARK PHYSICS

HEAVY QUARK PHYSICS

BY

DAYNAN BALL
ISABELLA HARRIS

KNOWLEDGE BAKERS

ISBN: 978-93-9001-396-8 (HB)

Published by : **Knowledge Bakers**
Pune, India
Email: info@knowledgebakers.com

PREFACE

Computation serves as the bedrock of modern technological advancement, underpinning innovations in fields as diverse as artificial intelligence, data science, cryptography, and software engineering. At its core, computation involves the manipulation of abstract symbols according to predefined rules, enabling the solution of cHeavy quarks, such as charm and bottom quarks, play a pivotal role in the realm of particle physics, offering unique insights into the fundamental forces and constituents of matter. The study of heavy quark physics not only contributes to our understanding of the Standard Model of particle physics but also provides a window into phenomena beyond the Standard Model, including the search for new particles and interactions. As we embark on this exploration of heavy quark physics, we delve into a realm where theoretical predictions, experimental observations, and computational techniques converge, uncovering the principles, phenomena, and frontiers that define this vibrant field of research.

This book, titled "Heavy Quark Physics," is crafted to provide a comprehensive overview of the key concepts, theories, experiments, and computational methods in the field of heavy quark physics. Through collaboration with leading theorists, experimentalists, and computational physicists in particle physics, we aim to offer readers a multidimensional and insightful journey through the intricacies of heavy quark dynamics, spectroscopy, decays, and interactions.

As editors of this volume, we have curated a collection of chapters that cover a wide range of topics in heavy quark physics, including quarkonium spectroscopy, heavy flavor decays, CP violation, rare decays, heavy quark effective theories, and lattice QCD calculations. Each chapter is structured to provide a theoretical foundation for understanding key concepts, followed by experimental results, theoretical predictions, and computational techniques to illustrate real-world applications and facilitate deeper comprehension.

Furthermore, this book seeks to highlight the interdisciplinary nature of heavy quark physics and its connections to other fields of physics, such as nuclear physics, astrophysics, and cosmology. By showcasing examples of how heavy quark physics contributes to our understanding of the universe, from the dynamics of quark-gluon plasma in heavy-ion collisions to the search for dark matter particles in astrophysical observations, we aim to inspire readers to explore new frontiers in particle physics and beyond.

As we embark on this enlightening journey through the world of heavy quark physics, we extend our gratitude to the contributors whose expertise and dedication have enriched this book. We hope that "Heavy Quark Physics" serves as a valuable resource for students, educators, researchers, and professionals in particle physics and related fields, empowering them to deepen their understanding of heavy quark dynamics and to contribute to the quest for new physics at the frontiers of knowledge.

Daynan Ball
Isabella Harris

Contents

1
Review

The standard model of strong, weak, and electromagnetic interactions is a relativistic quantum field theory that describes all known interactions of quarks and leptons. This chapter provides a quick review of features of the standard model that are relevant for heavy quark systems, and of basic field theory techniques such as the operator product expansion. It will also serve the purpose of defining some of the normalization conventions and notation to be used in the rest of the book.

1.1 The standard model

The standard model is a gauge theory based on the gauge group $SU(3) \times SU(2) \times U(1)$. The $SU(3)$ gauge group describes the strong color interactions among quarks, and the $SU(2) \times U(1)$ gauge group describes the electroweak interactions. At the present time three generations of quarks and leptons have been observed. The measured width of the Z boson does not permit a fourth generation with a massless (or light) neutrino. Many extensions of the minimal standard model have been proposed, and there is evidence in the present data for neutrino masses, which requires new physics beyond that in the minimal standard model. Low-energy supersymmetry, dynamical weak symmetry breaking, or something totally unexpected may be discovered at the next generation of high-energy particle accelerators.

The focus of this book is on understanding the physics of hadrons containing a bottom or charm quark. The technically difficult problem is understanding the role strong interactions play in determining the properties of these hadrons. For example, weak decays can be computed by using a low-energy effective weak Hamiltonian. Any new physics beyond the standard model can also be treated by using a local low-energy effective interaction, and the theoretical difficulties associated with evaluating hadronic matrix elements of this interaction are virtually identical to those for the weak interactions. For this reason, most of the discussion in this book will focus on the properties of heavy quark hadrons as computed in the standard model.

The matter fields in the minimal standard model are three families of spin-1/2 quarks and leptons, and a spin-zero Higgs boson, shown in Table 1.1. The index i on the Fermion fields is a family or generation index $i = 1, 2, 3$, and the subscripts L and R denote left- and right-handed fields, respectively,

$$\psi_L = P_L \psi, \qquad \psi_R = P_R \psi, \tag{1.1}$$

where P_L and P_R are the projection operators

$$P_L = \frac{1}{2}(1 - \gamma_5), \qquad P_R = \frac{1}{2}(1 + \gamma_5). \tag{1.2}$$

Q^i_L, u^i_R, d^i_R are the quark fields and L^i_L, e^i_R are the lepton fields. All the particles associated with the fields in Table 1.1 have been observed experimentally, except for the Higgs boson. The $SU(2) \times U(1)$ symmetry of the electroweak sector is not manifest at low energies. In the standard model, the $SU(2) \times U(1)$ symmetry is spontaneously broken by the vacuum expectation value of the Higgs doublet

Table 1.1. *Matter fields in the standard model*[a]

Field	$SU(3)$	$SU(2)$	$U(1)$	Lorentz
$Q^i_L = \begin{pmatrix} u^i_L \\ d^i_L \end{pmatrix}$	**3**	**2**	1/6	(1/2, 0)
u^i_R	**3**	**1**	2/3	(0, 1/2)
d^i_R	**3**	**1**	−1/3	(0, 1/2)
$L^i_L = \begin{pmatrix} \nu^i_L \\ e^i_L \end{pmatrix}$	**1**	**2**	−1/2	(1/2, 0)
e^i_R	**1**	**1**	−1	(0, 1/2)
$H = \begin{pmatrix} H^+ \\ H^0 \end{pmatrix}$	**1**	**2**	1/2	(0, 0)

[a] The index i labels the quark and lepton family. The dimensions of the $SU(3)$ and $SU(2)$ representations and their $U(1)$ charge are listed in the second, third, and fourth columns, respectively. The transformation properties of the fermion fields under the Lorentz group $SO(3, 1)$ are listed in the last column.

H. The spontaneous breakdown of $SU(2) \times U(1)$ gives mass to the $W^{\pm}$ and Z^0 gauge bosons. A single Higgs doublet is the simplest way to achieve the observed pattern of spontaneous symmetry breaking, but a more complicated scalar sector, such as two doublets, is possible.

The terms in the standard model Lagrangian density that involve only the Higgs doublet

$$H = \begin{pmatrix} H^+ \\ H^0 \end{pmatrix} \tag{1.3}$$

are

$$\mathcal{L}_{\text{Higgs}} = (D_\mu H)^\dagger (D^\mu H) - V(H), \tag{1.4}$$

where D_μ is the covariant derivative and $V(H)$ is the Higgs potential

$$V(H) = \frac{\lambda}{4}(H^\dagger H - v^2/2)^2. \tag{1.5}$$

The Higgs potential is minimized when $H^\dagger H = v^2/2$. The $SU(2) \times U(1)$ symmetry can be used to rotate a general vacuum expectation value into the standard form

$$\langle H \rangle = \begin{pmatrix} 0 \\ v/\sqrt{2} \end{pmatrix}, \tag{1.6}$$

where v is real and positive.

The generators of the $SU(2)$ gauge symmetry acting on the Higgs (i.e., fundamental) representation are

$$T^a = \sigma^a/2, \quad a = 1, 2, 3, \tag{1.7}$$

where the Pauli spin matrices are

$$\sigma^1 = \begin{pmatrix} 0 & 1 \\ 1 & 0 \end{pmatrix}, \quad \sigma^2 = \begin{pmatrix} 0 & -i \\ i & 0 \end{pmatrix}, \quad \sigma^3 = \begin{pmatrix} 1 & 0 \\ 0 & -1 \end{pmatrix}, \tag{1.8}$$

and the generators are normalized to $\text{Tr}\, T^a T^b = \delta^{ab}/2$. The $U(1)$ generator Y is called hypercharge and is equal to $1/2$ acting on the Higgs doublet (see Table 1.1). One linear combination of $SU(2) \times U(1)$ generators is left unbroken by the vacuum expectation value of the Higgs field H given in Eq. (1.6). This linear combination is the electric charge generator $Q = T^3 + Y$, where

$$Q = T^3 + Y = \begin{pmatrix} 1 & 0 \\ 0 & 0 \end{pmatrix}, \tag{1.9}$$

when acting on the Higgs representation. It is obvious from Eqs. (1.6) and (1.9) that

$$Q\langle H \rangle = 0, \tag{1.10}$$

so that electric charge is left unbroken. The $SU(3) \times SU(2) \times U(1)$ symmetry of the standard model is broken to $SU(3) \times U(1)_Q$ by the vacuum expectation value of H, where the unbroken electromagnetic $U(1)_Q$ is the linear combination of the original $U(1)$ hypercharge generator, Y, and the $SU(2)$ generator, T^3, given in Eq. (1.9).

Expanding H about its expectation value

$$H(x) = \begin{pmatrix} h^+(x) \\ v/\sqrt{2} + h^0(x) \end{pmatrix} \tag{1.11}$$

and substituting in Eq. (1.5) gives the Higgs potential

$$V(H) = \frac{\lambda}{4}(|h^+|^2 + |h^0|^2 + \sqrt{2}v \operatorname{Re} h^0)^2. \tag{1.12}$$

The fields h^+ and $\operatorname{Im} h^0$ are massless. This is an example of Goldstone's theorem. The potential has a continuous three-parameter family of degenerate vacua that are obtained from the reference vacuum in Eq. (1.6) by global $SU(2) \times U(1)$ transformations. [Of the four $SU(2) \times U(1)$ generators, one linear combination Q leaves the vacuum expectation value invariant, and so does not give a massless mode.] Field excitations along these degenerate directions cost no potential energy and so the fields h^+ and $\operatorname{Im} h^0$ are massless. There is one massive scalar that is destroyed by the (normalized) real scalar field $\sqrt{2} \operatorname{Re} h^0$. At tree level, its mass is

$$m_{\operatorname{Re} h^0} = \sqrt{\frac{\lambda}{2}}\, v. \tag{1.13}$$

Global $SU(2) \times U(1)$ transformations allow the space–time independent vacuum expectation value of H to be put into the form given in Eq. (1.6). Local $SU(2) \times U(1)$ transformations can be used to eliminate $h^+(x)$ and $\operatorname{Im} h^0(x)$ completely from the theory, and to write

$$H(x) = \begin{pmatrix} 0 \\ v/\sqrt{2} + \operatorname{Re} h^0(x) \end{pmatrix}. \tag{1.14}$$

This is the standard model in unitary gauge, in which the $W^\pm$ and Z bosons have explicit mass terms in the Lagrangian, as is shown below. In this gauge, the massless fields h^+ and $\operatorname{Im} h^0$ are eliminated, and so do not correspond to states in the spectrum of the theory.

The gauge covariant derivative acting on any field ψ is

$$D_\mu = \partial_\mu + igA_\mu^A T^A + ig_2 W_\mu^a T^a + ig_1 B_\mu Y, \tag{1.15}$$

where T^A, $A = 1, \ldots, 8$, are the eight color $SU(3)$ generators T^a, $a = 1, 2, 3$ are the weak $SU(2)$ generators, and Y is the $U(1)$ hypercharge generator. The generators are chosen to be in the representation of the field ψ on which the covariant derivative acts. The gauge bosons and coupling constants associated with these gauge groups are denoted A^A_μ, W^a_μ, and B_μ and g, g_2, and g_1, respectively. The kinetic term for the Higgs field contains a piece quadratic in the gauge fields when expanded about the Higgs vacuum expectation value using Eq. (1.11). The quadratic terms that produce a gauge-boson mass are

$$\mathcal{L}_{\substack{\text{gauge-boson}\\ \text{mass}}} = \frac{g_2^2 v^2}{8}(W^1 W^1 + W^2 W^2) + \frac{v^2}{8}(g_2 W^3 - g_1 B)^2, \tag{1.16}$$

where for simplicity of notation Lorentz indices are suppressed. The charged W-boson fields

$$W^{\pm} = \frac{W^1 \mp i W^2}{\sqrt{2}} \tag{1.17}$$

have mass

$$M_W = \frac{g_2 v}{2}. \tag{1.18}$$

It is convenient to introduce the weak mixing angle θ_W defined by

$$\sin\theta_W = \frac{g_1}{\sqrt{g_1^2 + g_2^2}}, \qquad \cos\theta_W = \frac{g_2}{\sqrt{g_1^2 + g_2^2}}. \tag{1.19}$$

The Z-boson field and photon field $\mathcal{A}$ are defined as linear combinations of the neutral gauge-boson fields W^3 and B,

$$\begin{aligned} Z &= \cos\theta_W W^3 - \sin\theta_W B, \\ \mathcal{A} &= \sin\theta_W W^3 + \cos\theta_W B. \end{aligned} \tag{1.20}$$

The Z boson has a mass at tree level

$$M_Z = \frac{\sqrt{g_1^2 + g_2^2}}{2} v = \frac{M_W}{\cos\theta_W}, \tag{1.21}$$

and the photon is massless.

The covariant derivative in Eq. (1.15) can be reexpressed in terms of the mass-eigenstate fields as

$$\begin{aligned} D_\mu = \partial_\mu &+ i g A^A_\mu T^A + i\frac{g_2}{\sqrt{2}}\left(W^+_\mu T^+ + W^-_\mu T^-\right) \\ &+ i\sqrt{g_1^2 + g_2^2}(T_3 - \sin^2\theta_W Q) Z_\mu + i g_2 \sin\theta_W Q \mathcal{A}_\mu, \end{aligned} \tag{1.22}$$

where $T^{\pm} = T^1 \pm i T^2$. The photon coupling constant in Eq. (1.22) leads to the relation between the electric charge e and the couplings $g_{1,2}$,

$$e = g_2 \sin\theta_W = \frac{g_2 g_1}{\sqrt{g_1^2 + g_2^2}}, \tag{1.23}$$

so the Z coupling constant $\sqrt{g_1^2 + g_2^2}$ in Eq. (1.22) is conventionally written as $e/(\sin\theta_W \cos\theta_W)$.

Outside of unitary gauge the H kinetic term also has a piece quadratic in the fields where the Goldstone bosons h^+, $\mathrm{Im}\, h^0$ mix with the longitudinal parts of the massive gauge bosons. This mixing piece can be removed by adding to the Lagrange density the 't Hooft gauge fixing term

$$\begin{aligned}\mathcal{L}_{\substack{\text{gauge}\\ \text{fix}}} &= -\frac{1}{2\xi}\sum_a \left[\partial^\mu W_\mu^a + ig_2\xi(\langle H\rangle^\dagger T^a H - H^\dagger T^a \langle H\rangle)\right]^2 \\ &\quad -\frac{1}{2\xi}\left[\partial^\mu B_\mu + ig_1\xi(\langle H\rangle^\dagger Y H - H^\dagger Y \langle H\rangle)\right]^2, \end{aligned} \tag{1.24}$$

which gives the Lagrangian in R_ξ gauge, where ξ is an arbitrary parameter. The fields $h^\pm$ and $\mathrm{Im}\, h^0$ have mass terms proportional to the gauge fixing constant ξ. In Feynman gauge $\xi = 1$ (the easiest for doing calculations), these masses are the same as those of the $W^\pm$ and Z. $\mathrm{Im}\, h^0$ and $h^\pm$ are not physical degrees of freedom since in unitary gauge $\xi \to \infty$ their masses are infinite and they decouple from the theory.

$SU(3) \times SU(2) \times U(1)$ gauge invariance prevents bare mass terms for the quarks and leptons from appearing in the Lagrange density. The quarks and leptons get mass because of their Yukawa couplings to the Higgs doublet,

$$\mathcal{L}_{\text{Yukawa}} = g_u^{ij}\, \bar{u}_R^i H^T \epsilon Q_L^j - g_d^{ij}\, \bar{d}_R^i H^\dagger Q_L^j - g_e^{ij}\, \bar{e}_R^i H^\dagger L_L^j + \text{h.c.} \tag{1.25}$$

where h.c. denotes Hermitian conjugate. Here repeated indices i, j are summed and the antisymmetric matrix ϵ is given by

$$\epsilon = \begin{pmatrix} 0 & 1 \\ -1 & 0 \end{pmatrix}. \tag{1.26}$$

Color indices and spinor indices are suppressed in Eq. (1.25). Since H has a vacuum expectation value, the Yukawa couplings in Eq. (1.25) give rise to the 3×3 quark and lepton mass matrices

$$\mathcal{M}_u = v g_u/\sqrt{2}, \quad \mathcal{M}_d = v g_d/\sqrt{2}, \quad \text{and } \mathcal{M}_e = v g_e/\sqrt{2}. \tag{1.27}$$

Neutrinos do not get mass from the Yukawa interactions in Eq. (1.25), since there is no right-handed neutrino field.

Any matrix M can be brought into diagonal form by separate unitary transformations on the left and right, $M \to LDR^\dagger$, where L and R are unitary, and D is real, diagonal and nonnegative. One can make separate unitary transformations on the left- and right-handed quark and lepton fields, while leaving the kinetic energy terms for the quarks, $\bar{Q}_L^i i\not{\partial}\, Q_L^i$, $\bar{u}_R^i i\not{\partial}\, u_R^i$, and $\bar{d}_R^i i\not{\partial}\, d_R^i$, and also those for the leptons, invariant. The unitary transformations are

$$\begin{aligned} u_L &= \mathcal{U}(u, L)\, u'_L, & u_R &= \mathcal{U}(u, R)\, u'_R, \\ d_L &= \mathcal{U}(d, L)\, d'_L, & d_R &= \mathcal{U}(d, R)\, d'_R, \\ e_L &= \mathcal{U}(e, L)\, e'_L, & e_R &= \mathcal{U}(e, R)\, e'_R. \end{aligned} \tag{1.28}$$

Here u, d, and e are three-component column vectors (in flavor space) for the quarks and leptons, and the primed fields represent the corresponding mass eigenstates. The transformation matrices $\mathcal{U}$ are 3×3 unitary matrices, which are chosen to diagonalize the mass matrices

$$\mathcal{U}(u, R)^\dagger \mathcal{M}_u \mathcal{U}(u, L) = \begin{pmatrix} m_u & 0 & 0 \\ 0 & m_c & 0 \\ 0 & 0 & m_t \end{pmatrix}, \tag{1.29}$$

$$\mathcal{U}(d, R)^\dagger \mathcal{M}_d \mathcal{U}(d, L) = \begin{pmatrix} m_d & 0 & 0 \\ 0 & m_s & 0 \\ 0 & 0 & m_b \end{pmatrix}, \tag{1.30}$$

and

$$\mathcal{U}(e, R)^\dagger \mathcal{M}_e \mathcal{U}(e, L) = \begin{pmatrix} m_e & 0 & 0 \\ 0 & m_\mu & 0 \\ 0 & 0 & m_\tau \end{pmatrix}. \tag{1.31}$$

Diagonalizing the quark mass matrices in Eqs. (1.29) and (1.30) requires different transformations of the u_L and d_L fields, which are part of the same $SU(2)$ doublet Q_L. The original quark doublet can be rewritten as

$$\begin{pmatrix} u_L \\ d_L \end{pmatrix} = \begin{pmatrix} \mathcal{U}(u, L)\, u'_L \\ \mathcal{U}(d, L)\, d'_L \end{pmatrix} = \mathcal{U}(u, L) \begin{pmatrix} u'_L \\ V d'_L \end{pmatrix}, \tag{1.32}$$

where the Cabibbo-Kobayashi-Maskawa (CKM) mixing matrix V is defined by

$$V = \mathcal{U}(u, L)^\dagger \mathcal{U}(d, L). \tag{1.33}$$

It is convenient to reexpress the standard model Lagrangian in terms of the primed mass-eigenstate fields. The unitary matrices in Eq. (1.32) leave the quark kinetic terms unchanged. The Z and $\mathcal{A}$ couplings are also unaffected, so there are no flavor-changing neutral currents in the Lagrangian at tree level. The W couplings are left unchanged by $\mathcal{U}(u, L)$, but not by V, so that

$$\frac{g_2}{\sqrt{2}} W^+ \bar{u}_L \gamma^\mu d_L = \frac{g_2}{\sqrt{2}} W^+ \bar{u}'_L \gamma^\mu V d'_L. \tag{1.34}$$

As a result there are flavor-changing charged currents at tree level.

The CKM matrix V is a 3×3 unitary matrix, and so is completely specified by nine real parameters. Some of these can be eliminated by making phase redefinitions of the quark fields. The u and d quark mass matrices are unchanged if one makes independent phase rotations on the six quarks, provided the same phase is used for the left- and right-handed quarks of a given flavor. An overall equal phase rotation on all the quarks leaves the CKM matrix unchanged, but the remaining five rotations can be used to eliminate five parameters, so that V is written in terms of four parameters. The original Kobayashi-Maskawa parameterization of V is

$$V = \begin{pmatrix} c_1 & s_1 c_3 & s_1 s_3 \\ -s_1 c_2 & c_1 c_2 c_3 - s_2 s_3 e^{i\delta} & c_1 c_2 s_3 + s_2 c_3 e^{i\delta} \\ -s_1 s_2 & c_1 s_2 c_3 + c_2 s_3 e^{i\delta} & c_1 s_2 s_3 - c_2 c_3 e^{i\delta} \end{pmatrix}, \tag{1.35}$$

where $c_i \equiv \cos\theta_i$, and $s_i \equiv \sin\theta_i$ for $i = 1, 2, 3$. The angles θ_1, θ_2, and θ_3 can be chosen to lie in the first quadrant, where their sines and cosines are positive. Experimentally it is known that these angles are quite small. The CKM matrix is real if $\delta = 0$, so that $\delta \neq 0$ is a signal of CP violation in the weak interactions. It describes the unitary transformation between the mass-eigenstate basis $d^{i\prime}$, and the weak interaction eigenstate basis d^i. The standard notation for the mass-eigenstate fields is $u'^1 = u$, $u'^2 = c$, $u'^3 = t$, $d'^1 = d$, $d'^2 = s$, $d'^3 = b$.

So far we have only considered the left-handed quark couplings to the gauge bosons. For the right-handed quarks there are no W-boson interactions in the standard model, and in the primed mass-eigenstate basis the couplings of the Z, photon, and color gauge bosons are flavor diagonal. The analysis for leptons is similar to that for quarks, with one notable difference – because the neutrinos are massless, one can choose to make the same unitary transformation on the left-handed charged leptons and neutrinos. The analog of the CKM matrix in the lepton sector can be chosen to be the unit matrix, and the leptons can be chosen to be simultaneously mass and weak eigenstates. We adopt the notation $\nu'^1 = \nu_e$, $\nu'^2 = \nu_\mu$, $\nu'^3 = \nu_\tau$, $e'^1 = e$, $e'^2 = \mu$, $e'^3 = \tau$. From now on, we will use the mass-eigenstate basis for labeling the quark and lepton fields.

1.2 Loops

Loop diagrams in the standard model have divergences from the high-momentum (ultraviolet) region of the momentum integrals. These divergences are interpreted by a renormalization procedure; the theory is regulated in some way and terms that diverge as the regulator is removed are absorbed into the definitions of the couplings and masses. Theories in which all divergences in physical quantities (e.g., S-matrix elements) can be removed in this way using a finite number of counterterms are called renormalizable. In the unitary gauge, $\xi \to \infty$, the standard model is manifestly unitary (i.e., only physical degrees of freedom propagate because the "ghost" Higgs associated with $h^\pm$ and $\mathrm{Im}\, h^0$ have infinite mass). The vector-boson propagator

$$-i\,\frac{g_{\mu\nu} - k_\mu k_\nu / M_{W,Z}^2}{k^2 - M_{W,Z}^2} \tag{1.36}$$

is finite as $k \to \infty$, and naive power counting suggests that the standard model is not renormalizable. In the Feynman gauge, $\xi = 1$, the vector-boson propagator is

$$-i\,\frac{g_{\mu\nu}}{k^2 - M_{W,Z}^2}, \tag{1.37}$$

which falls off as $1/k^2$, and naive power counting shows that the standard model is renormalizable. The potentially disastrous divergences that occur in the unitary gauge must cancel. However, unitarity is not manifest in the Feynman gauge because the unphysical degrees of freedom associated with $h^\pm$ and $\operatorname{Im} h^0$ are included as intermediate states in Feynman diagrams. The standard model is manifestly unitary in one gauge and manifestly renormalizable in another. Gauge invariance assures us that the theory is both unitary and renormalizable.

In this book we will regularize Feynman diagrams by using dimensional regularization. Diagrams are calculated in $n = 4 - \epsilon$ dimensions, and the ultraviolet divergences that occur in four dimensions appear as factors of $1/\epsilon$, as $\epsilon \to 0$.

To review how dimensional regularization works, consider the quantum electrodynamics (QED) Lagrangian

$$\mathcal{L}_{\mathrm{QED}} = -\frac{1}{4} F_{\mu\nu}^{(0)} F^{(0)\mu\nu} + i\bar{\psi}^{(0)} \gamma^\mu \left(\partial_\mu - i e^{(0)} \mathcal{A}_\mu^{(0)}\right) \psi^{(0)} - m_e^{(0)} \bar{\psi}^{(0)} \psi^{(0)}, \tag{1.38}$$

which is part of the standard model Lagrangian. The superscript (0) is used to denote a bare quantity. Here

$$F_{\mu\nu}^{(0)} = \partial_\mu \mathcal{A}_\nu^{(0)} - \partial_\nu \mathcal{A}_\mu^{(0)} \tag{1.39}$$

is the bare electromagnetic field strength tensor. In n dimensions, the action

$$S_{\mathrm{QED}} = \int d^n x\, \mathcal{L}_{\mathrm{QED}} \tag{1.40}$$

is dimensionless, since $e^{i S_{\mathrm{QED}}}$ is the measure in the Feynman path integral (we use units where $\hbar = c = 1$). It follows that the dimensions of the fields, the coupling constant $e^{(0)}$, and the electron mass, $m_e^{(0)}$, are

$$\begin{aligned}
\left[\mathcal{A}^{(0)}\right] &= (n-2)/2 = 1 - \epsilon/2, \\
\left[\psi^{(0)}\right] &= (n-1)/2 = 3/2 - \epsilon/2, \\
\left[e^{(0)}\right] &= (4-n)/2 = \epsilon/2, \\
\left[m_e^{(0)}\right] &= 1.
\end{aligned} \tag{1.41}$$

The bare fields are related to the renormalized fields by

$$
\begin{aligned}
\mathcal{A}_\mu &= \frac{1}{\sqrt{Z_A}}\mathcal{A}_\mu^{(0)}, \\
\psi &= \frac{1}{\sqrt{Z_\psi}}\psi^{(0)}, \\
e &= \frac{1}{Z_e}\mu^{-\epsilon/2}e^{(0)}, \\
m_e &= \frac{1}{Z_m}m_e^{(0)}.
\end{aligned}
\tag{1.42}
$$

The factor of $\mu^{-\epsilon/2}$ is included in the relation between the bare and renormalized electric couplings so that the renormalized coupling is dimensionless. Here μ is a parameter with dimensions of mass and is called the subtraction point or renormalization scale of dimensional regularization. In terms of these renormalized quantities the Lagrange density is

$$
\begin{aligned}
\mathcal{L}_{\text{QED}} &= -\frac{1}{4}Z_A F_{\mu\nu}F^{\mu\nu} + iZ_\psi \bar{\psi}\gamma^\mu\left(\partial_\mu - i\mu^{\epsilon/2}Z_e\sqrt{Z_A}e\mathcal{A}_\mu\right)\psi \\
&\quad - Z_m Z_\psi m_e \bar{\psi}\psi, \\
&= -\frac{1}{4}F_{\mu\nu}F^{\mu\nu} + i\bar{\psi}\gamma^\mu\left(\partial_\mu - i\mu^{\epsilon/2}e\mathcal{A}_\mu\right)\psi - m_e\bar{\psi}\psi + \text{ counterterms.}
\end{aligned}
\tag{1.43}
$$

It is straightforward to compute the renormalization constants $Z_{A,\psi,e,m}$ by using the formula for one-loop integrals in dimensional regularization,

$$
\begin{aligned}
&\int \frac{d^n q}{(2\pi)^n}\frac{(q^2)^\alpha}{(q^2-M^2)^\beta} \\
&\quad = \frac{i}{2^n\pi^{n/2}}(-1)^{\alpha+\beta}(M^2)^{\alpha-\beta+n/2}\frac{\Gamma(\alpha+n/2)\Gamma(\beta-\alpha-n/2)}{\Gamma(n/2)\Gamma(\beta)},
\end{aligned}
\tag{1.44}
$$

and the Feynman trick for combining denominators,

$$
\begin{aligned}
\frac{1}{a_1^{m_1}\cdots a_n^{m_n}} &= \frac{\Gamma(M)}{\Gamma(m_1)\cdots\Gamma(m_n)} \\
&\quad \times \int_0^1 dx_1 x_1^{m_1-1}\cdots\int_0^1 dx_n x_n^{m_n-1}\frac{\delta\left(1-\sum_{i=1}^n x_i\right)}{(x_1a_1+\cdots+x_na_n)^M},
\end{aligned}
\tag{1.45}
$$

where

$$
M = \sum_{i=1}^n m_i.
$$

The Z's are determined by the condition that time-ordered products of renormalized fields (i.e., Green's functions) be finite when expressed in terms of the renormalized coupling and mass. This condition still leaves considerable freedom in how the Z's are chosen. The precise way that the Z's are chosen is called the subtraction scheme. The Z's can be chosen to have the form

$$Z = 1 + \sum_{p=1}^{\infty} \frac{Z_p(e)}{\epsilon^p}, \tag{1.46}$$

where the $Z_p(e)$ are independent of ϵ. This choice is called minimal subtraction (MS) because only the poles in ϵ are subtracted and no additional finite pieces are put into the Z's. We will use the $\overline{\text{MS}}$ scheme, which is minimal subtraction followed by the rescaling $\mu^2 \to \mu^2 e^\gamma/4\pi$, where $\gamma = 0.577\ldots$ is Euler's constant.

The photon wavefunction renormalization Z_A to order e^2 can be determined by computing the photon–photon correlation function. There are two pieces to this order; the first is a tree-level contribution from the counterterm

$$-\frac{1}{4}(Z_A - 1)F_{\mu\nu}F^{\mu\nu}. \tag{1.47}$$

After truncating the external photon propagators, it gives

$$i(Z_A - 1)(p_\mu p_\nu - p^2 g_{\mu\nu}), \tag{1.48}$$

where p is the photon four momentum. The second contribution is from the one-loop diagram Fig. 1.1,

$$(-1)\,(ie)^2\,\mu^\epsilon \int \frac{d^n q}{(2\pi)^n} \frac{\text{Tr}[\gamma_\mu i(\not{q} + \not{p} + m_e)\gamma_\nu i\,(\not{q} + m_e)]}{[(q+p)^2 - m_e^2][q^2 - m_e^2]}. \tag{1.49}$$

The factor of (-1) arises from the closed fermion loop. The renormalization constant only depends on the $1/\epsilon$ pole, so the γ matrix algebra can be performed in four dimensions. Expanding

$$\mu^\epsilon = 1 + \epsilon\,\ln\mu + \cdots, \tag{1.50}$$

one sees that μ^ϵ can be set to unity for the infinite part of the diagram, and finite

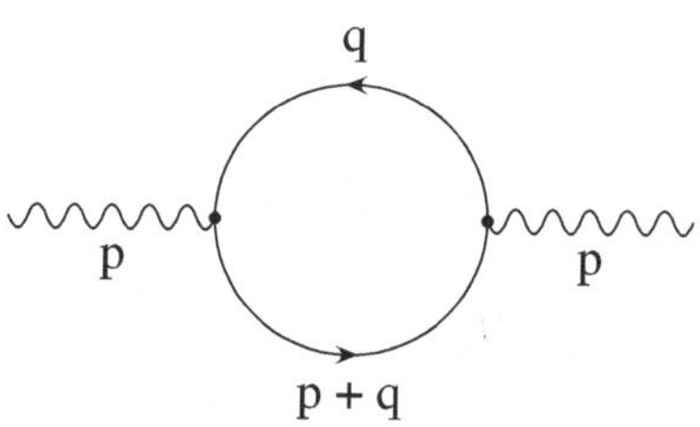

Fig. 1.1. One-loop vacuum polarization contribution to the photon propagator.

parts only depend logarithmically on μ. The denominators are combined using Eq. (1.45):

$$\frac{1}{\left[(q+p)^2 - m_e^2\right]\left[q^2 - m_e^2\right]} = \int_0^1 dx \frac{1}{\left[q^2 + 2xq\cdot p + p^2 x - m_e^2\right]^2}. \tag{1.51}$$

Making the change of variables $k = q + px$ gives

$$-4e^2 \int_0^1 dx \int \frac{d^n k}{(2\pi)^n} \frac{1}{\left[k^2 + p^2 x(1-x) - m_e^2\right]^2} \times \left[2k_\mu k_\nu - \left(k^2 - m_e^2\right) g_{\mu\nu} - 2x(1-x) p_\mu p_\nu + p^2 x(1-x) g_{\mu\nu}\right]. \tag{1.52}$$

Terms odd in k vanish upon integration and have been dropped. Evaluating the k integral using Eq. (1.44), keeping only the part proportional to $1/\epsilon$ (using $\Gamma(\epsilon/2) = 2/\epsilon + \cdots$), and doing the x integral gives the divergent part of the one-loop contribution:

$$\frac{i}{16\pi^2\epsilon}\left(\frac{8e^2}{3}\right)\left(p_\mu p_\nu - p^2 g_{\mu\nu}\right). \tag{1.53}$$

For the photon two-point correlation function to be finite as $\epsilon \to 0$, the sum of Eqs. (1.53) and (1.48) must be finite. One therefore chooses

$$Z_A = 1 - \frac{8}{3}\left(\frac{e^2}{16\pi^2\epsilon}\right). \tag{1.54}$$

The wave-function renormalization constant Z_ψ for the electron field ψ is obtained from the electron propagator. The counterterms

$$(Z_\psi - 1)\bar\psi i \not\partial \psi - (Z_m Z_\psi - 1) m_e \bar\psi \psi \tag{1.55}$$

contribute

$$i(Z_\psi - 1)\not p - i(Z_m Z_\psi - 1) m_e \tag{1.56}$$

to the propagator. In the Feynman gauge, the one-loop diagram Fig. 1.2 is

$$\mu^\epsilon (ie)^2 \int \frac{d^n q}{(2\pi)^n} \gamma_\nu i \frac{\not p + \not q + m_e}{(p+q)^2 - m_e^2} \gamma_\mu \frac{(-i) g^{\mu\nu}}{q^2}. \tag{1.57}$$

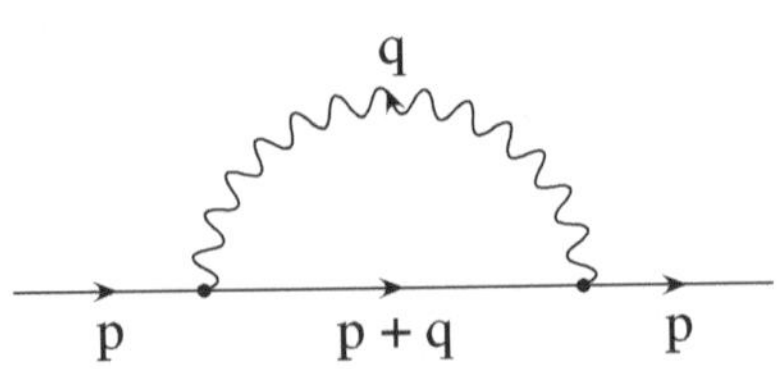

Fig. 1.2. One-loop correction to the electron propagator.

Combining denominators and shifting the momentum integration as in the previous case gives

$$2e^2 \int \frac{d^n k}{(2\pi)^n} \int_0^1 dx \frac{-2m_e + \not{p}(1-x)}{\left[k^2 - m_e^2 x + p^2 x(1-x)\right]^2}. \tag{1.58}$$

Performing the k integration by using Eq. (1.44) and then the x integration gives

$$\frac{i}{16\pi^2\epsilon}(4e^2)\left(-2m_e + \frac{1}{2}\not{p}\right) \tag{1.59}$$

for the divergent contribution. The electron propagator is finite if

$$Z_\psi = 1 - 2\left(\frac{e^2}{16\pi^2\epsilon}\right) \tag{1.60}$$

and

$$Z_m = 1 - 6\left(\frac{e^2}{16\pi^2\epsilon}\right) \tag{1.61}$$

in the Feynman gauge.

The remaining renormalization factor Z_e can be determined by computing the $\psi\bar{\psi}A$ three-point function to order e^2. The Feynman graph that has to be computed is the vertex renormalization graph of Fig. 1.3. The counterterm is

$$Z_e = 1 + \frac{4}{3}\left(\frac{e^2}{16\pi^2\epsilon}\right). \tag{1.62}$$

Note that $Z_e = 1/\sqrt{Z_A}$ to order e^2.

The relation between the bare and renormalized couplings at order e^2 is

$$e^{(0)} = \mu^{\epsilon/2} e Z_e = \mu^{\epsilon/2} e \left[1 + \frac{4}{3}\left(\frac{e^2}{16\pi^2\epsilon}\right)\right], \tag{1.63}$$

using Eqs. (1.62) and (1.42). The bare fields, coupling, and mass are independent of the subtraction point μ, which is an arbitrary quantity with dimensions of mass introduced so that the renormalized coupling is dimensionless. Since the bare

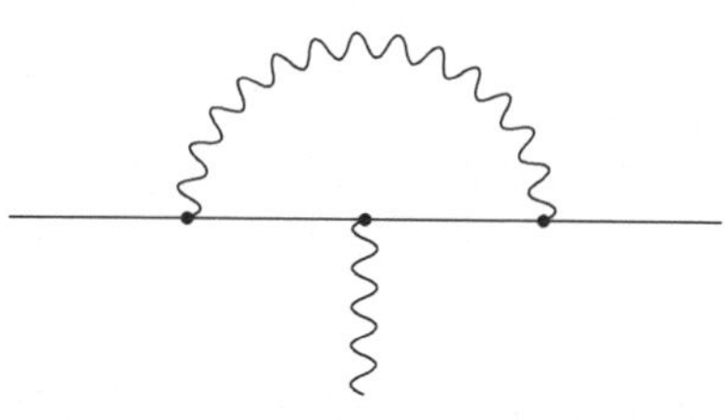

Fig. 1.3. One-loop vertex correction.

coupling constant is independent of μ,

$$0 = \mu\frac{\mathrm{d}}{\mathrm{d}\mu}e^{(0)} = \mu\frac{\mathrm{d}}{\mathrm{d}\mu}\mu^{\epsilon/2}eZ_e = \mu^{\epsilon/2}eZ_e\left[\frac{\epsilon}{2} + \frac{1}{e}\beta\,(e) + \frac{\mu}{Z_e}\frac{\mathrm{d}Z_e}{\mathrm{d}\mu}\right], \tag{1.64}$$

where the β function is defined by

$$\beta(e) = \mu\frac{\mathrm{d}e}{\mathrm{d}\mu}. \tag{1.65}$$

This gives

$$\beta(e) = -\frac{\epsilon}{2}e - e\frac{\mathrm{d}\ln Z_e}{\mathrm{d}\ln\mu}. \tag{1.66}$$

Using Eq. (1.62),

$$\begin{aligned}\frac{\mathrm{d}\ln Z_e}{\mathrm{d}\ln\mu} &= \left(\frac{4}{3}\right)\frac{1}{16\pi^2\epsilon}\mu\frac{\mathrm{d}}{\mathrm{d}\mu}e^2 + \cdots \\ &= -\frac{e^2}{12\pi^2} + \cdots,\end{aligned} \tag{1.67}$$

where the ellipses denote terms of higher order in e^2. The one-loop β function is

$$\beta(e) = -\frac{\epsilon}{2}e + \frac{e^3}{12\pi^2} + \cdots, \tag{1.68}$$

which is finite as $\epsilon \to 0$,

$$\beta(e) = \frac{e^3}{12\pi^2} + \cdots . \tag{1.69}$$

The β function gives the μ dependence of the renormalized coupling e. Here μ is an arbitrary scale parameter, so physical quantities do not depend on μ. However, some choices for μ are more convenient than others for computations. Consider the cross section for $\sigma(e^+e^- \to \text{anything})$ at a center of mass energy squared, $s = (p_{e^+} + p_{e^-})^2 \gg m_e^2$. In QED this cross section is finite as $m_e \to 0$ and so for large s we neglect m_e. The cross section has a power series expansion in the coupling $e(\mu)$, and it is independent of the subtraction point μ. The implicit μ dependence in the coupling is canceled by an explicit μ dependence in the Feynman diagrams. (One can see this by computing, e.g., the finite parts of Figs. 1.1–1.3.) Typically one finds that terms in the perturbation series have the form $[\alpha\,(\mu)\,/4\pi]^n \ln^m s/\mu^2$, with $m \le n$, where

$$\alpha\,(\mu) = \frac{e^2(\mu)}{4\pi} \tag{1.70}$$

is the (scale-dependent) fine structure constant. If s/μ^2 is not of the order of unity, the logarithms can get large and cause a breakdown of perturbation theory.

One usually chooses $\mu^2 \sim s$, which "minimizes" the higher-order terms in the perturbation expansion *that have not been computed*. With this choice of μ, one expects that perturbation theory is an expansion in $\alpha(\sqrt{s})/4\pi$.

When perturbation theory is valid we can use Eqs. (1.65) and (1.69) to solve explicitly for the dependence of the coupling on μ at one loop:

$$\frac{1}{e^2(\mu_2)} = \frac{1}{e^2(\mu_1)} - \frac{1}{12\pi^2}\ln\left(\frac{\mu_2^2}{\mu_1^2}\right). \tag{1.71}$$

The β function in Eq. (1.69) is positive, so e increases as μ increases, as can be seen explicitly from the solution in Eq. (1.71).

1.3 Composite operators

Composite operators involve products of fields at the same space–time point. Consider, for example, the bare mass operator

$$S^{(0)} = \bar\psi^{(0)}\psi^{(0)}(x). \tag{1.72}$$

Green's functions with an insertion of $S^{(0)}$ are usually divergent. An additional operator renormalization (beyond wave-function renormalization) is required to make the Green's functions finite. The renormalized operator S is

$$S = \frac{1}{Z_S}S^{(0)} = \frac{1}{Z_S}\bar\psi^{(0)}\psi^{(0)} = \frac{Z_\psi}{Z_S}\bar\psi\psi, \tag{1.73}$$

where Z_S is the additional operator renormalization. The operator $S = \bar\psi\psi +$ counterterms is conventionally denoted by just $\bar\psi\psi$, with the counterterms implicit. Green's functions with insertions of S are finite in perturbation theory.

The renormalization factor Z_S can be computed from the three-point function of the time-ordered product of ψ, $\bar\psi$, and S. It is simpler to use the one-particle irreducible Green's function Γ rather than the full Green's function G to compute Z_S. The counterterm contribution to the one-particle irreducible Green's function is

$$\frac{Z_\psi}{Z_S} - 1. \tag{1.74}$$

The one-loop contribution to Γ is shown in Fig. 1.4. The operator S contains no derivatives (and Z_S is mass independent in the $\overline{\text{MS}}$ scheme), so Z_S can be determined by evaluating Fig. 1.4 at zero external momentum (and neglecting the electron mass), giving

$$\mu^\epsilon (ie)^2 \int \frac{d^nq}{(2\pi)^n}\gamma^\alpha \frac{i\not{q}}{q^2}\frac{i\not{q}}{q^2}\gamma^\beta \frac{(-i)g_{\alpha\beta}}{q^2} = -4ie^2\int\frac{d^nq}{(2\pi)^n}\frac{1}{(q^2)^2} + \cdots, \tag{1.75}$$

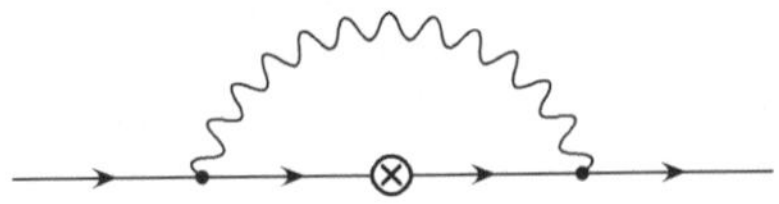

Fig. 1.4. One-loop graph with an insertion of a fermion-bilinear composite operator (denoted by $\otimes$) such as $\bar{\psi}\psi$.

where the ellipsis denote terms finite as $\epsilon \to 0$. Note that neglecting external momenta and the electron mass has produced an infrared (i.e., low momentum) divergence. Regulating this with a mass m by replacing q^2 in the denominator with $(q^2 - m^2)$ gives

$$\frac{8e^2}{16\pi^2\epsilon}, \tag{1.76}$$

for the ultraviolet divergent part of Eq. (1.75). Adding Eqs. (1.74) and (1.76) together and using Eq. (1.60), we find that the $1/\epsilon$ divergence cancels, provided

$$Z_S = 1 + 6\left(\frac{e^2}{16\pi^2\epsilon}\right). \tag{1.77}$$

The anomalous dimension of the composite operator S is defined by

$$\gamma_S = \mu\frac{\mathrm{d}\ln Z_S}{\mathrm{d}\mu} \tag{1.78}$$

so that

$$\gamma_S = -\frac{6e^2}{16\pi^2}. \tag{1.79}$$

Similar calculations can be performed for the vector and axial vector currents $\bar{\psi}\gamma_\mu\psi$ and $\bar{\psi}\gamma_\mu\gamma_5\psi$, and one finds $Z_V = Z_A = 1$, so that the currents are not renormalized and their anomalous dimensions vanish at one loop. Note that $Z = 1$ means that the infinite part of Fig. 1.4 is canceled by wave-function renormalization, not that Fig. 1.4 is finite. The result $Z = 1$ arises because for $m_e = 0$ both the axial and vector currents are conserved and the zero-component of these currents (integrated over all space) are charges $Q_{A,V}$ with commutation relations of the form

$$[Q_V, \psi] = -\psi, \tag{1.80}$$

for example. A conserved charge Q cannot be multiplicatively renormalized since that would spoil such commutation relations. In dimensional regularization with minimal subtraction, electron mass effects cannot induce a renormalization for the axial current because the renormalization factors are independent of particle masses. This is an example of a general result that "soft" symmetry breaking effects, i.e., symmetry breaking terms with operator dimensions less than four, do not affect renormalization in the $\overline{\text{MS}}$ scheme.

The axial current is not conserved at one loop because of the axial anomaly. The divergence of the axial current is proportional to the dimension-four operator $F\tilde{F}$, so that symmetry breaking because of the anomaly is not soft. It produces an anomalous dimension for the axial current at two loops.

We have considered a particularly simple example in which the operator S was multiplicatively renormalized, since there are no other gauge invariant local operators with the same quantum numbers. In general, one can have many different operators O_i with the same quantum numbers, and one needs a renormalization matrix,

$$O_i^{(0)} = Z_{ij} O_j. \tag{1.81}$$

This is referred to as operator mixing. In the $\overline{\text{MS}}$ scheme, Z_{ij} is dimensionless, so operators can only mix with other operators of the same dimension. This greatly simplifies the analysis of operator mixing. In a general mass-dependent scheme, operators can also mix with operators of lower dimension.

1.4 Quantum chromodynamics and chiral symmetry

The portion of the standard model that describes the strong interactions of quarks and gluons is called quantum chromodynamics (QCD). The QCD Lagrange density including for the moment only the "light" u, d, and s quark flavors is

$$\mathcal{L}_{\text{QCD}} = -\frac{1}{4} G^A_{\mu\nu} G^{A\mu\nu} + \bar{q}(i\not{D} - m_q)q + \text{counterterms}, \tag{1.82}$$

where q is the triplet of light quarks

$$q = \begin{pmatrix} u \\ d \\ s \end{pmatrix}, \tag{1.83}$$

and m_q is the quark mass matrix

$$m_q = \begin{pmatrix} m_u & 0 & 0 \\ 0 & m_d & 0 \\ 0 & 0 & m_s \end{pmatrix}. \tag{1.84}$$

Here $D_\mu = \partial_\mu + igA^A_\mu T^A$ is the $SU(3)$ color covariant derivative and $G^A_{\mu\nu}$ is the gluon field strength tensor,

$$G^A_{\mu\nu} = \partial_\mu A^A_\nu - \partial_\nu A^A_\mu - gf^{ABC} A^B_\mu A^C_\nu, \tag{1.85}$$

where the structure constants f^{ABC} are defined by $[T^A, T^B] = if^{ABC}T^C$. The QCD renormalization factors can be calculated at order g^2 in a manner similar

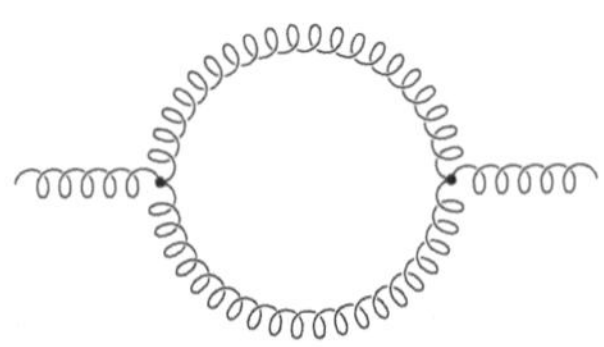

Fig. 1.5. One-loop gluon contribution to the vacuum polarization.

to that for QED. For example, quark wave-function and mass renormalization Z_q and Z_m are given by Fig. 1.2 with the photon replaced by a gluon. They can be obtained from the QED result by replacing e^2 by $g^2 T^A T^A$, where $T^A T^A = (4/3)\mathbb{1}$ for quarks in QCD. In the Feynman gauge, the order g^2 wave-function and mass renormalization factors are

$$\sqrt{Z_q} = 1 - \frac{g^2}{12\pi^2 \epsilon}, \qquad Z_m = 1 - \frac{g^2}{2\pi^2 \epsilon}. \tag{1.86}$$

A major difference between QCD and QED occurs in the coupling constant renormalization. The β function for QCD is

$$\beta(g) = -\frac{g^3}{16\pi^2}\left(11 - \frac{2}{3}N_q\right) + \mathcal{O}(g^5), \tag{1.87}$$

where N_q is the number of quark flavors. The quark contribution to the β function can be computed from Fig. 1.1 with the photon replaced by a gluon. It is obtained from the QED calculation by the replacement $e^2 \to N_q g^2/2$, since $\mathrm{Tr}\, T^A T^B = \delta^{AB}/2$ for each quark flavor in the loop. The other term in the β function is from gluon self-interactions, as in Fig. 1.5, and is not present in an Abelian gauge theory such as QED. The QCD β function is negative, as long as the number of quark flavors N_q is less than 16, so the QCD fine structure constant

$$\alpha_s(\mu) = \frac{g^2(\mu)}{4\pi} \tag{1.88}$$

becomes smaller at larger μ, a phenomenon known as asymptotic freedom. At high energies, the coupling constant is small, and QCD perturbation theory should be reliable. We can explicitly solve for the μ-dependence of α_s just as in QED:

$$\alpha_s(\mu_2) = \frac{1}{\left[1/\alpha_s(\mu_1) + \beta_0 \ln\left(\mu_2^2/\mu_1^2\right)\right]}, \tag{1.89}$$

where β_0 is proportional to the first term in the QCD β function,

$$\beta_0 = \left(\frac{33 - 2N_q}{12\pi}\right). \tag{1.90}$$

Equation (1.89) is valid as long as μ_1 and μ_2 are large enough that the order g^5 terms in Eq. (1.87) can be neglected, i.e., as long as $\alpha_s(\mu_1)$ and $\alpha_s(\mu_2)$ are both small. It is convenient to introduce a subtraction-point independent constant Λ_{QCD} with dimensions of mass, defined by

$$\Lambda_{\text{QCD}} = \mu e^{-1/[2\beta_0 \alpha_s(\mu)]}. \tag{1.91}$$

Then our expression for the strong interaction fine structure constant becomes

$$\alpha_s(\mu) = \frac{12\pi}{(33 - 2N_q)\ln\left(\mu^2/\Lambda_{\text{QCD}}^2\right)}. \tag{1.92}$$

Equation (1.92) suggests that the QCD coupling constant diverges as $\mu \to \Lambda_{\text{QCD}}$. Of course, this expression for α_s ceases to be valid when α_s gets large. Nevertheless, one can still view Λ_{QCD} as the scale at which QCD becomes strongly coupled so that perturbation theory breaks down and nonperturbative effects become important. Experimentally, Λ_{QCD} is $\sim$200 MeV, and it sets the scale for nonperturbative strong interaction effects. One expects hadron masses such as the ρ meson mass to be dimensionless multiples of Λ_{QCD}. It is believed that QCD is a confining theory at long distances, i.e., the spectrum of physical states consists of color singlet states called hadrons; there are no colored hadrons. Bosonic hadrons are called mesons and fermionic hadrons are called baryons. The simplest ways to form color singlet combinations of the quark fields are $\bar{q}^\alpha q_\alpha$ and $\varepsilon^{\alpha\beta\gamma} q_\alpha q_\beta q_\gamma$.

The u, d, and s quark masses are small compared with the scale Λ_{QCD} of nonperturbative strong interaction physics, and so it is useful to consider an approximation to QCD in which the masses of these light quarks are set to zero, and to do perturbation theory in m_q about this limit. The limit $m_q \to 0$ is known as the chiral limit, because the light quark Lagrangian

$$\mathcal{L}_{\substack{\text{light}\\ \text{quarks}}} = \bar{q}\, i\not{D}\, q = \bar{q}_L\, i\not{D}\, q_L + \bar{q}_R\, i\not{D}\, q_R \tag{1.93}$$

has an $SU(3)_L \times SU(3)_R$ chiral symmetry

$$q_L \to L\, q_L \qquad q_R \to R\, q_R, \tag{1.94}$$

$[L \in SU(3)_L,\ R \in SU(3)_R]$ under which the right- and left-handed quark fields transform differently. The Lagrange density in Eq. (1.93) also has a baryon number $U(1)$ symmetry where the left- and right-handed quarks transform by a common phase, and an axial $U(1)$ where all the left-handed quarks transform by a phase and all the right-handed quarks transform by the opposite phase. Although these axial $U(1)$ transformations leave the Lagrange density invariant, they change the measure in the path integral, an effect known as the axial anomaly. Hence, the axial $U(1)$ is not a symmetry of QCD.

The chiral $SU(3)_L \times SU(3)_R$ symmetry of massless three-flavor QCD is spontaneously broken by the vacuum expectation value of quark bilinears

$$\left\langle \bar{q}_R^j q_L^k \right\rangle = v\, \delta^{kj}, \tag{1.95}$$

where v is of order Λ_{QCD}^3. [Here v should not be confused with the Higgs vacuum expectation value.] The indices j and k are flavor indices, $q^1 = u, q^2 = d, q^3 = s$, and color indices are suppressed. If we make a $SU(3)_L \times SU(3)_R$ transformation $q \to q'$,

$$\langle \bar{q}'^j_R q'^k_L \rangle = v(LR^\dagger)^{kj}. \tag{1.96}$$

Transformations with $L = R$ leave the vacuum expectation value unchanged. Thus the nonperturbative strong interaction dynamics spontaneously breaks the $SU(3)_L \times SU(3)_R$ chiral symmetry to its diagonal subgroup $SU(3)_V$. The eight broken $SU(3)_L \times SU(3)_R$ generators transform the composite field $\bar{q}^j_R q^k_L$ along symmetry directions, and so leave the potential energy unchanged. Fluctuations in field space along these eight directions are eight massless Goldstone bosons. We can describe the Goldstone boson fields by a 3×3 special unitary matrix $\Sigma(x)$, which represents the possible low-energy long-wavelength excitations of $\bar{q}_R q_L$. Here $v\Sigma_{kj}(x) \sim \bar{q}^j_R(x) q^k_L(x)$ gives the local orientation of the quark condensate. Σ has vacuum expectation value $\langle \Sigma \rangle = \mathbb{1}$. Under $SU(3)_L \times SU(3)_R$ transformations,

$$\Sigma \to L\,\Sigma\,R^\dagger. \tag{1.97}$$

The low-momentum strong interactions of the Goldstone bosons are described by an effective Lagrangian for $\Sigma(x)$ that is invariant under the chiral symmetry transformation in Eq. (1.97). The most general Lagrangian is

$$\mathcal{L}_{\text{eff}} = \frac{f^2}{8} \operatorname{Tr} \partial^\mu \Sigma\, \partial_\mu \Sigma^\dagger + \text{ higher derivative terms}, \tag{1.98}$$

where f is a constant with dimensions of mass. There are no terms without any derivatives since $\operatorname{Tr} \Sigma \Sigma^\dagger = 3$. At a low enough momentum the effects of the higher derivative terms can be neglected since they are suppressed by powers of $p^2_{\text{typ}}/\Lambda^2_{\text{CSB}}$, where p_{typ} is a typical momentum and Λ_{CSB} is the scale associated with chiral symmetry breaking, $\Lambda_{\text{CSB}} \sim 1$ GeV.

The field $\Sigma(x)$ is an $SU(3)$ matrix and it can be written as the exponential

$$\Sigma = \exp\left(\frac{2iM}{f}\right), \tag{1.99}$$

of M, a traceless 3×3 Hermitian matrix. Under the unbroken $SU(3)_V$ subgroup $(L = R = V)$, $\Sigma \to V\Sigma V^\dagger$, which implies that $M \to VMV^\dagger$, i.e., M transforms as the adjoint representation. M can be written out explicitly in terms of eight Goldstone boson fields:

$$M = \begin{pmatrix} \pi^0/\sqrt{2} + \eta/\sqrt{6} & \pi^+ & K^+ \\ \pi^- & -\pi^0/\sqrt{2} + \eta/\sqrt{6} & K^0 \\ K^- & \bar{K}^0 & -2\eta/\sqrt{6} \end{pmatrix}. \tag{1.100}$$

The factor of $2/f$ is inserted in Eq. (1.99) so that the Lagrangian in Eq. (1.98) gives kinetic-energy terms for the Goldstone bosons with the standard normalization.

In the QCD Lagrangian the light quark mass terms,

$$\mathcal{L}_{\text{mass}} = \bar{q}_L m_q q_R + \text{h.c.}, \tag{1.101}$$

transform under chiral $SU(3)_L \times SU(3)_R$ as $(\bar{\mathbf{3}}_L, \mathbf{3}_R) + (\mathbf{3}_L, \bar{\mathbf{3}}_R)$. We can include the effects of quark masses (to first order) on the strong interactions of the pseudo-Goldstone bosons, π, K, and η, by adding terms linear in m_q to Eq. (1.98) that transform in this way. Equivalently we can view the quark mass matrix itself as transforming like $m_q \to L m_q R^\dagger$ under $SU(3)_L \times SU(3)_R$. Then the Lagrange density in Eq. (1.101) is invariant under chiral $SU(3)_L \times SU(3)_R$. With this transformation rule for m_q, we include the effects of quark masses in the strong interactions of the π, K, and η by adding to Eq. (1.98) terms linear in m_q and $m_q^\dagger$ that are invariant under $SU(3)_L \times SU(3)_R$. This gives

$$\mathcal{L}_{\text{eff}} = \frac{f^2}{8}\text{Tr}\,\partial^\mu\Sigma\,\partial_\mu\Sigma^\dagger + v\,\text{Tr}(m_q^\dagger\Sigma + m_q\Sigma^\dagger) + \cdots. \tag{1.102}$$

The ellipses in Eq. (1.102) represent terms with more derivatives or more insertions of the light quark mass matrix. The quark mass terms in the Lagrange density in Eq. (1.102) give masses to the Goldstone bosons

$$\begin{aligned} m_{\pi^\pm}^2 &= \frac{4v}{f^2}(m_u + m_d), \\ m_{K^\pm}^2 &= \frac{4v}{f^2}(m_u + m_s), \\ m_{K^0}^2 = m_{\bar{K}^0}^2 &= \frac{4v}{f^2}(m_d + m_s), \end{aligned} \tag{1.103}$$

and hence the π, K, and η are referred to as pseudo-Goldstone bosons. The kaon masses are much larger than the pion masses, implying that $m_s \gg m_{u,d}$. For the $\eta - \pi^0$ system there is a mass-squared matrix with elements

$$\begin{aligned} m_{\pi^0\pi^0}^2 &= \frac{4v}{f^2}(m_u + m_d), \\ m_{\eta\pi^0}^2 = m_{\pi^0\eta}^2 &= \frac{4v}{\sqrt{3}f^2}(m_u - m_d), \\ m_{\eta\eta}^2 &= \frac{4v}{3f^2}(4m_s + m_u + m_d). \end{aligned} \tag{1.104}$$

Because $m_s \gg m_{u,d}$, the off-diagonal terms are small compared with $m_{\eta\eta}^2$. Hence, up to corrections suppressed by $(m_u - m_d)^2/m_s^2$,

$$m_{\pi^0}^2 \simeq \frac{4v}{f^2}(m_u + m_d) \tag{1.105}$$

and

$$m_\eta^2 \simeq \frac{4v}{3f^2}(4m_s + m_u + m_d)\,. \tag{1.106}$$

It is interesting to note that the neutral pion mass is near the charged pion masses, not because m_u/m_d is near unity, but rather because $m_u - m_d$ is small compared with m_s. A more detailed study of mass relations, including electromagnetic corrections, leads to the expectation that $m_u/m_d \simeq 1/2$.

The chiral Lagrangian in Eq. (1.102) contains two parameters, v with dimensions of (mass)^3 and f with dimensions of mass. Since the quark masses always appear in conjunction with v it is not possible using the effective Lagrangian in Eq. (1.102) to determine the quark masses themselves. The effective theory describing the low-momentum interactions of the pseudo-Goldstone bosons only determines the ratios of quark masses, since v cancels out.

Equation (1.102) is an effective Lagrangian that describes the low-momentum interactions of the pseudo-Goldstone bosons. One can use the effective theory to compute scattering processes, such as $\pi - \pi$ scattering. Expanding out Σ in terms of the meson fields, one finds that the $\mathrm{Tr}\, \partial_\mu \Sigma\, \partial^\mu \Sigma^\dagger$ part of the Lagrangian has the four-meson interaction term,

$$\frac{1}{6f^2}\mathrm{Tr}[M, \partial_\mu M][M, \partial^\mu M]. \tag{1.107}$$

Its tree-level matrix element (shown in Fig. 1.6) gives a contribution to the π–π scattering amplitude of the form

$$\mathcal{M} \sim \frac{p_{\text{typ}}^2}{f^2}, \tag{1.108}$$

where p_{typ} is a typical momentum. The amplitude is of the order of p_{typ}^2 since the vertex contains two derivatives. The mass terms also give a contribution of this form if we set $p_{\text{typ}}^2 \sim m_\pi^2$. The contributions of higher derivative operators

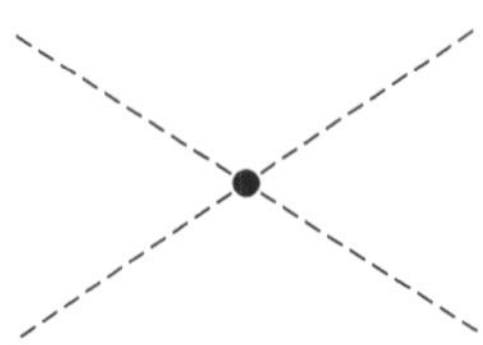

Fig. 1.6. Tree-level contribution to π–π scattering.

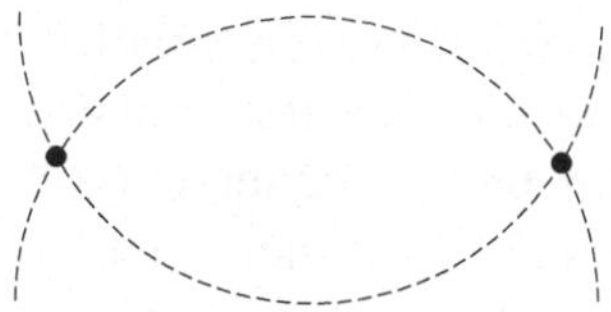

Fig. 1.7. One-loop contribution to π–π scattering.

in the chiral Lagrangian are suppressed by more factors of the small momentum p_{typ}.

What about loop diagrams? There are one-loop diagrams with two insertions of the $\pi\pi\pi\pi$ vertex, such as Fig. 1.7. Each vertex gives a factor of p^2/f^2, the two meson propagators give a factor of $1/p^4$, and the loop integration gives a factor of p^4. The resulting amplitude in the $\overline{\text{MS}}$ scheme is

$$\mathcal{M} \sim \frac{p_{\text{typ}}^4}{16\pi^2 f^4} \ln\left(p_{\text{typ}}^2/\mu^2\right). \tag{1.109}$$

The factor of p_{typ}^4 in the numerator is required by dimensional analysis, since there is a factor of f^4 in the denominator and the subtraction point μ, which also has dimensions of mass, only occurs in the argument of logarithms. The $16\pi^2$ in the denominator typically occurs in the evaluation of one-loop diagrams. The one-loop diagram gives a contribution of the same order in the momentum expansion as operators in the chiral Lagrangian with four derivatives (or two insertions of the quark mass matrix). The total amplitude at order p^4 is the sum of one-loop diagrams containing order p^2 vertices and of tree graphs from the p^4 terms in the Lagrangian. The total p^4 amplitude is μ independent; the μ dependence in Eq. (1.109) is canceled by μ dependence in the coefficients of the p^4 terms in the Lagrangian.

The pattern we have just observed holds in general. More loops give a contribution of the same order as a term in the Lagrangian with more derivatives. One can show that a graph with L loops, and n_k insertions of vertices of order p^k, produces an amplitude of order p^D, where (see Problem 6)

$$D = 2 + 2L + \sum_k (k-2)\, n_k. \tag{1.110}$$

Thus each loop increases D by two, and each insertion of a vertex of order p^k increases D by $k-2$. Note that $k-2 \geq 0$, since the Lagrangian starts at order p^2, so that each term in Eq. (1.110) is positive. Loop corrections and higher derivative operators are of comparable importance when the mass scale Λ_{CSB} that suppresses higher derivative operators is approximately equal to $4\pi f$. The computation of pseudo-Goldstone scattering amplitudes in a momentum expansion using an effective Lagrangian is known as chiral perturbation theory.

Although the u, d, and s quark masses are small, the spectrum of QCD suggests that the theory contains quasi-particles that transform like u, d, and s under the unbroken $SU(3)_V$ group but have a larger mass of approximately 350 MeV. These quasi-particles are called constituent quarks, and the hadronic spectrum is consistent at least qualitatively with spectra calculated from nonrelativistic potential models for the interactions of constituent quarks.

1.5 Integrating out heavy quarks

The top, bottom, and charm quark masses are $m_t \simeq 175$ GeV, $m_b \simeq 4.8$ GeV, and $m_c \simeq 1.4$ GeV. For processes that occur at energies well below the masses of these quarks, it is appropriate to go over to an effective theory of the strong interactions where these heavy quarks are integrated out of the theory and no longer occur as explicit degrees of freedom in the Lagrangian. The effects of Feynman diagrams with a virtual heavy quark Q are taken into account by non-renormalizable operators suppressed by factors of $1/m_Q$, and through shifts in the coupling constants of renormalizable terms in the effective Lagrangian. For definiteness, imagine integrating out the top quark and making a transition from the six-quark theory of the strong interactions to an effective five-quark theory. The strong coupling in the original theory with six flavors will be denoted by $g^{(6)}$, and in the effective five-quark theory by $g^{(5)}$. The relation between the two couplings is determined by ensuring that the scattering amplitudes computed in the five- and six-quark theories are the same. The general form of the relation is $g^{(5)}(\mu) = g^{(5)}[m_t/\mu, g^{(6)}(\mu)]$, since the g's are dimensionless. The power series expansion of $g^{(5)}$ in powers of $g^{(6)}$ has coefficients that, for μ very different from m_t, contain large logarithms of m_t^2/μ^2. If instead we pick $\mu = m_t$, $g^{(5)}$ has a power series expansion in $g^{(6)}(m_t)$ with coefficients that are not enhanced by any large logarithms. At tree level, $g^{(5)}(\mu) = g^{(6)}(\mu)$, so one expects

$$g^{(5)}(m_t) = g^{(6)}(m_t)\left\{1 + \mathcal{O}\left[\alpha_s^{(6)}(m_t)\right]\right\}. \tag{1.111}$$

An explicit calculation shows that the one-loop term in this equation vanishes, so the first nontrivial contribution is at two loops. The strong coupling in the effective theory with n quarks is written as in Eq. (1.92), where the value of Λ_{QCD} now depends on which particular effective theory is being used (i.e., $\Lambda_{\text{QCD}} \to \Lambda_{\text{QCD}}^{(n)}$). Equation (1.111) implies that, at leading order, the coupling constants are continuous at $\mu = m_t$. Combining this with Eq. (1.92), we find that

$$\Lambda_{\text{QCD}}^{(5)} = \Lambda_{\text{QCD}}^{(6)}\left(\frac{m_t}{\Lambda_{\text{QCD}}^{(6)}}\right)^{2/23}. \tag{1.112}$$

Integrating out the bottom and charm quarks to go over to effective four- and three-quark theories gives

$$\Lambda_{\rm QCD}^{(4)} = \Lambda_{\rm QCD}^{(5)} \left(\frac{m_b}{\Lambda_{\rm QCD}^{(5)}} \right)^{2/25}, \tag{1.113}$$

$$\Lambda_{\rm QCD}^{(3)} = \Lambda_{\rm QCD}^{(4)} \left(\frac{m_c}{\Lambda_{\rm QCD}^{(4)}} \right)^{2/27}. \tag{1.114}$$

Equations (1.112)–(1.114) determine the most important influence of virtual heavy quarks on low-energy physics. For example, the proton mass m_p is generated by nonperturbative dynamics in the effective three-quark theory so $m_p \propto \Lambda_{\rm QCD}^{(3)}$, where the constant of proportionality is independent of the heavy quark masses. Imagining that the value of the strong coupling is fixed at some very-high-energy scale (e.g., the unification scale), Eqs. (1.112)–(1.114) give the dependence of the proton mass on the heavy quark masses. For example, doubling the charm quark mass increases the proton mass by the factor $2^{2/27} \simeq 1.05$.

1.6 Effective Hamiltonians for weak decays

The strong and electromagnetic interactions conserve quark and lepton flavor, so many particles can only decay by means of the weak interactions. The simplest example of such a decay is the weak decay of a muon, $\mu \to e \nu_\mu \bar{\nu}_e$. This decay is a purely leptonic process, since it does not involve any quark fields. The lowest-order graph for this decay has a single W boson exchanged, as shown in Fig. 1.8. The tree-level amplitude for the decay is

$$\begin{aligned} \mathcal{M}(\mu \to e \nu_\mu \bar{\nu}_e) &= \left(\frac{g_2}{\sqrt{2}} \right)^2 [\bar{u}(p_{\nu_\mu}) \gamma_\alpha P_L u(p_\mu)][\bar{u}(p_e) \gamma_\beta P_L v(p_{\nu_e})] \\ &\times \frac{1}{[(p_\mu - p_{\nu_\mu})^2 - M_W^2]} \left[g^{\alpha\beta} - \frac{(p_\mu - p_{\nu_\mu})^\alpha (p_\mu - p_{\nu_\mu})^\beta}{M_W^2} \right], \end{aligned} \tag{1.115}$$

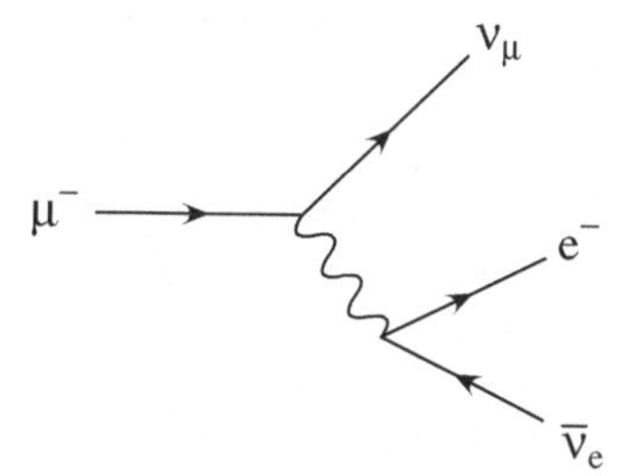

Fig. 1.8. Lowest-order diagram for μ decay.

where g_2 is the weak $SU(2)$ coupling constant, and the W propagator has been written in the unitary gauge. The muon mass is much smaller than the W-boson mass M_W, so the momenta of all the leptons involved in μ decay are much smaller than M_W. As a result, we can approximate the denominator of the W-boson propagator, $(p_\mu - p_{\nu_\mu})^2 - M_W^2$, by $-M_W^2$ and neglect the factor of $(p_\mu - p_{\nu_\mu})^\alpha (p_\mu - p_{\nu_\mu})^\beta / M_W^2$ in the numerator of the W-boson propagator. This approximation simplifies the decay amplitude to

$$\mathcal{M}(\mu \to e \nu_\mu \bar{\nu}_e) \simeq -\frac{4G_F}{\sqrt{2}} \left[\bar{u}(p_{\nu_\mu}) \gamma_\alpha P_L u(p_\mu)\right]\left[\bar{u}(p_e) \gamma^\alpha P_L v(p_{\nu_e})\right], \qquad (1.116)$$

where the Fermi constant G_F is defined by

$$\frac{G_F}{\sqrt{2}} = \frac{g_2^2}{8M_W^2}. \qquad (1.117)$$

The decay amplitude Eq. (1.116) is the same as that produced by the tree-level matrix element of the local effective Hamiltonian:

$$H_W = -\mathcal{L}_W = \frac{4G_F}{\sqrt{2}} \left[\bar{\nu}_\mu \gamma_\alpha P_L \mu\right]\left[\bar{e} \gamma^\alpha P_L \nu_e\right]. \qquad (1.118)$$

It is simpler to use an effective Hamiltonian description of the weak interactions in computing weak decay amplitudes at energies much smaller than M_W and M_Z, particularly if one wants to compute radiative corrections to decay amplitudes.

Electromagnetic loop corrections to the $\mu \to e \bar{\nu}_e \nu_\mu$ decay amplitude go partly into matrix elements of the Hamiltonian in Eq. (1.118) and partly into modifying the Hamiltonian itself. The corrections to the Hamiltonian are calculated by comparing amplitudes in the full theory with the W boson present as a dynamical field to amplitudes in the effective theory with the W boson removed. These corrections come from regions of loop momenta of order M_W, since the effective Hamiltonian has been chosen to correctly reproduce the full Hamiltonian for momenta much smaller than M_W. For this reason, the electron and muon masses occur as $m_{e,\mu}/M_W$ in the effective Hamiltonian, and they can be neglected at leading order. They are, of course, very important for the matrix elements of the effective Hamiltonian.

Neglecting the electron and muon masses, we know that the Hamiltonian must be of the form in Eq. (1.118). In this limit electromagnetic corrections do not change chirality and so $[\bar{\nu}_\mu \gamma_\alpha P_L \mu][\bar{e} \gamma^\alpha P_L \nu_e]$ and $[\bar{\nu}_\mu \gamma_\alpha P_L \nu_e][\bar{e} \gamma^\alpha P_L \mu]$ are the only possible dimension-six operators that can occur. Terms with three gamma matrices between the fermion fields can be reduced to single gamma matrices by using the identity

$$\gamma_\alpha \gamma_\beta \gamma_\nu = g_{\alpha\beta} \gamma_\nu + g_{\beta\nu} \gamma_\alpha - g_{\alpha\nu} \gamma_\beta - i \epsilon_{\alpha\beta\nu\eta} \gamma^\eta \gamma_5, \qquad (1.119)$$

where the sign convention is $\epsilon_{0123} = 1$. Higher dimension operators are negligible, being suppressed by powers of $1/M_W$. The Fierz operator identity,

$$[\bar{\psi}_1\gamma_\alpha P_L\psi_2][\bar{\psi}_3\gamma^\alpha P_L\psi_4] = [\bar{\psi}_1\gamma_\alpha P_L\psi_4][\bar{\psi}_3\gamma^\alpha P_L\psi_2], \qquad (1.120)$$

allows one to replace $[\bar{\nu}_\mu\gamma_\alpha P_L\nu_e][\bar{e}\gamma^\alpha P_L\mu]$ by $[\bar{\nu}_\mu\gamma_\alpha P_L\mu][\bar{e}\gamma^\alpha P_L\nu_e]$. So beyond tree level the effective Hamiltonian is modified to

$$H_W = \frac{4G_F}{\sqrt{2}} C\left[\frac{M_W}{\mu}, \alpha(\mu)\right][\bar{\nu}_\mu\gamma^\alpha P_L\mu][\bar{e}\gamma_\alpha P_L\nu_e], \qquad (1.121)$$

where μ is the subtraction point, and α is the electromagnetic fine structure constant. The only modification due to radiative corrections is the coefficient C, which is unity at tree level. Loop corrections at $\mu = M_W$ with virtual loop momenta of order M_W determine the deviation of the coefficient C from unity, so one expects

$$C[1, \alpha(M_W)] = 1 + \mathcal{O}[\alpha(M_W)]. \qquad (1.122)$$

Any dependence of the matrix elements of the four-fermion operator $[\bar{\nu}_\mu\gamma_\alpha P_L \mu][\bar{e}\gamma^\alpha P_L\nu_e]$ on the subtraction point is canceled by the μ dependence of C, so that physical quantities such as decay rates do not depend on μ. If the Hamiltonian above is used to calculate the muon decay rate, with $\mu = M_W$ naively one would think that there are large logarithms of (m_μ^2/M_W^2) in the perturbative expansion of the matrix elements of the Hamiltonian. In fact we know that C is μ-independent and hence such logarithms do not occur. A simple explanation of this fact follows using the Fierz identity in Eq. (1.120), which allows us to rewrite the effective Hamiltonian in the form $[\bar{\nu}_\mu\gamma_\alpha P_L\nu_e][\bar{e}\gamma^\alpha P_L\mu]$. The neutrino fields do not interact electromagnetically, so the only renormalization is that of $\bar{e}\gamma^\alpha P_L\mu$. In the limit $m_e = m_\mu = 0$, $\bar{e}\gamma^\alpha P_L\mu$ is a conserved current and does not get renormalized.

The electromagnetic coupling α is so small that even when it is multiplied by large logarithms, perturbation theory is usually adequate. However, this is not the case for the strong interactions. For an example in which such logarithms are important and must be summed, consider the effective Hamiltonian for nonleptonic $b \to c$ decays at tree level,

$$H_W^{(\Delta c = 1)} = \frac{4G_F}{\sqrt{2}} V_{cb}V_{ud}^*[\bar{c}^\alpha\gamma_\mu P_L b_\alpha][\bar{d}^\beta\gamma^\mu P_L u_\beta]. \qquad (1.123)$$

In Eq. (1.123) α and β are color indices and repeated indices are summed. There is a contribution to the effective Hamiltonian for nonleptonic $b \to c$ decays where the d quark is replaced by a s quark. It has a coefficient that is suppressed by $|V_{us}/V_{ud}| \approx 0.2$ compared to Eq. (1.123). This "Cabibbo suppressed" contribution is neglected here. Also, we are focusing on $\Delta c = 1$ decays. There are nonleptonic decays where, at tree level, the final state has both a c and $\bar{c}$ quark. For these decays the coefficient in the effective Hamiltonian $H_W^{(\Delta c = 0)}$ is not smaller than Eq. (1.123).

Strong interaction loop corrections change the form of the Hamiltonian for $b \to c$ decays. An argument similar to that used for μ decay shows that there are two possible terms in the $\Delta c = 1$ effective Hamiltonian,

$$H_W = \frac{4G_F}{\sqrt{2}} V_{cb} V_{ud}^* \left\{ C_1 \left[\frac{M_W}{\mu}, \alpha_s(\mu) \right] O_1(\mu) + C_2 \left[\frac{M_W}{\mu}, \alpha_s(\mu) \right] O_2(\mu) \right\}, \tag{1.124}$$

where

$$\begin{aligned} O_1(\mu) &= [\bar{c}^\alpha \gamma_\mu P_L b_\alpha][\bar{d}^\beta \gamma^\mu P_L u_\beta], \\ O_2(\mu) &= [\bar{c}^\beta \gamma_\mu P_L b_\alpha][\bar{d}^\alpha \gamma^\mu P_L u_\beta]. \end{aligned} \tag{1.125}$$

The coefficients $C_{1,2}$ are determined by comparing Feynman diagrams in the effective theory with the W-boson integrated out with analogous diagrams in the full theory. At $\mu = M_W$ we have from Eq. (1.123) that

$$\begin{aligned} C_1[1, \alpha_s(M_W)] &= 1 + \mathcal{O}[\alpha_s(M_W)], \\ C_2[1, \alpha_s(M_W)] &= 0 + \mathcal{O}[\alpha_s(M_W)]. \end{aligned} \tag{1.126}$$

Subtraction-point dependence of the operators $O_{1,2}$ cancels that in the coefficients $C_{1,2}$. Here $O_{1,2}$ are local four-quark operators, and they must be renormalized to render their matrix elements finite. The relationship between bare and renormalized operators has the form

$$O_i^{(0)} = Z_{ij} O_j, \tag{1.127}$$

where $i, j = \{1, 2\}$ and the repeated index j is summed over. Since the bare operator is μ independent,

$$0 = \mu \frac{\mathrm{d}}{\mathrm{d}\mu} O_i^{(0)}(\mu) = \left(\mu \frac{\mathrm{d}}{\mathrm{d}\mu} Z_{ij} \right) O_j + Z_{ij} \left(\mu \frac{\mathrm{d}}{\mathrm{d}\mu} O_j \right), \tag{1.128}$$

which implies that

$$\mu \frac{\mathrm{d}}{\mathrm{d}\mu} O_j = -\gamma_{ji} O_i(\mu), \tag{1.129}$$

where

$$\gamma_{ji} = Z_{jk}^{-1} \left(\mu \frac{\mathrm{d}}{\mathrm{d}\mu} Z_{ki} \right). \tag{1.130}$$

Here $\gamma_{ij}(g)$ is called the anomalous dimension matrix. It can be calculated order by order in the coupling constant from the Z's. The subtraction-point independence of the weak Hamiltonian implies that

$$0 = \mu \frac{\mathrm{d}}{\mathrm{d}\mu} H_W = \mu \frac{\mathrm{d}}{\mathrm{d}\mu} (C_j O_j), \tag{1.131}$$

yielding

$$\left(\mu\frac{\mathrm{d}}{\mathrm{d}\mu}C_j\right)O_j - C_j\gamma_{ji}O_i = 0. \tag{1.132}$$

Since the operators $O_{1,2}$ are independent we conclude that

$$\mu\frac{\mathrm{d}}{\mathrm{d}\mu}C_i = \gamma_{ji}C_j. \tag{1.133}$$

The solution to this differential equation is

$$C_i\left[\frac{M_W}{\mu}, \alpha_s(\mu)\right] = P\exp\left[\int_{g(M_W)}^{g(\mu)}\frac{\gamma^T(g)}{\beta(g)}dg\right]_{ij} C_j[1, \alpha_s(M_W)]. \tag{1.134}$$

Here P denotes "coupling constant ordering" of the anomalous dimension matrices in the exponent, and γ^T is the transpose of γ.

It is straightforward to calculate the anomalous dimension matrix for $O_{1,2}$. At one loop, it is

$$\gamma(g) = \frac{g^2}{8\pi^2}\begin{pmatrix} -1 & 3 \\ 3 & -1 \end{pmatrix}. \tag{1.135}$$

It is convenient to diagonalize this matrix by forming the linear combinations of operators

$$O_\pm = O_1 \pm O_2. \tag{1.136}$$

Using the Fierz identity in Eq. (1.120), it is evident that O_+ is symmetric under interchange of the d and c quark fields, whereas O_- is antisymmetric. Under an $SU(2)$ flavor group under which the d and c quark fields form a doublet, O_- is a singlet and O_+ is a triplet. The c, d mass difference breaks this flavor symmetry. Quark masses do not affect the renormalization constants Z_{ij}, so mixing between O_+ and O_- is forbidden by this symmetry. In terms of $O_\pm$ the effective weak Hamiltonian is

$$H_W = \frac{4G_F}{\sqrt{2}}V_{cb}V_{ud}^*\left\{C_+\left[\frac{M_W}{\mu}, \alpha_s(\mu)\right]O_+(\mu) + C_-\left[\frac{M_W}{\mu}, \alpha_s(\mu)\right]O_-(\mu)\right\}, \tag{1.137}$$

where

$$C_\pm[1, \alpha_s(M_W)] = \frac{1}{2} + \mathcal{O}[\alpha_s(M_W)]. \tag{1.138}$$

At any other subtraction point

$$C_\pm\left[\frac{M_W}{\mu}, \alpha_s(\mu)\right] = \exp\left[\int_{g(M_W)}^{g(\mu)}\frac{\gamma_\pm(g)}{\beta(g)}dg\right]C_\pm[1, \alpha_s(M_W)], \tag{1.139}$$

where

$$\gamma_+(g) = \frac{g^2}{4\pi^2} + \mathcal{O}(g^4),$$
$$\gamma_-(g) = -\frac{g^2}{2\pi^2} + \mathcal{O}(g^4), \tag{1.140}$$

and $\beta(g)$ is given by Eq. (1.87). Provided $\mu \gg \Lambda_{\rm QCD}$, the strong coupling $\alpha_s(\mu)$ is small over the range of integration in Eq. (1.139) and higher-order terms in g can be neglected in $\gamma_\pm$ and β. This gives

$$C_\pm\left[\frac{M_W}{\mu}, \alpha_s(\mu)\right] = \frac{1}{2}\left[\frac{\alpha_s(M_W)}{\alpha_s(\mu)}\right]^{a_\pm}, \tag{1.141}$$

where

$$a_+ = \frac{6}{33 - 2N_q}, \qquad a_- = -\frac{12}{33 - 2N_q}. \tag{1.142}$$

Expressing $\alpha_s(M_W)$ in terms of $\alpha_s(\mu)$ using Eq. (1.89), the perturbative power series expansions of $C_\pm$ have the form

$$\frac{1}{2} + a_1\, \alpha_s(\mu) \ln(M_W/\mu) + a_2\, \alpha_s^2(\mu) \ln^2(M_W/\mu) + \cdots. \tag{1.143}$$

The expression for $C_\pm$ in Eq. (1.143) sums all leading logarithms of the form $\alpha_s^n(\mu) \ln^n(M_W/\mu)$, neglecting subleading logarithms of order $\alpha_s^n \ln^{n-1}(M_W/\mu)$. The series of subleading logarithms can be summed by using two-loop renormalization group equations, and so on. The subtraction-point dependence in the coefficients $C_\pm$ cancels that in the matrix elements of the operators $O_\pm$ so any value of μ can be used. However, if $p_{\rm typ}$ is the typical momentum in a nonleptonic decay, the matrix elements of $O_\pm$ will contain large logarithms of $(\mu^2/p_{\rm typ}^2)$, for μ very different from $p_{\rm typ}$. Roughly these logarithms come from integrations over momenta in the region between $p_{\rm typ}$ and μ. They are summed by scaling the coefficients down from the subtraction point M_W to one of order $p_{\rm typ}$, which moves the logarithms from the matrix elements of $O_\pm$ to the coefficients $C_\pm$.

The exponents $a_\pm$ in Eq. (1.142) depend on the number of quark flavors N_q. It is convenient to integrate out the top quark at the same time as the W boson so that $N_q = 5$. For inclusive weak decay of a hadron containing a b quark, the typical momenta of the decay products are of the order of the b-quark mass, and the large logarithms of $(M_W/m_b)^2$ are summed by evaluating the coefficients $C_\pm$ at $\mu = m_b$. In this case,

$$C_+(m_b) = 0.42, \qquad C_-(m_b) = 0.70, \tag{1.144}$$

using $\alpha_s(M_W) = 0.12$ and $\alpha_s(m_b) = 0.22$.

1.7 The pion decay constant

Weak pion decay $\pi^- \to \mu\bar{\nu}_\mu$ determines the value for the parameter f that occurs in the chiral Lagrangian for pion strong interactions in Eq. (1.98). Neglecting electromagnetic corrections, the effective Hamiltonian for $\pi^- \to \mu\bar{\nu}_\mu$ decay is

$$H_{\text{eff}} = \frac{4G_F}{\sqrt{2}} V_{ud}[\bar{u}\gamma_\alpha P_L d][\bar{\mu}\gamma^\alpha P_L \nu_\mu]. \tag{1.145}$$

Here color indices on the quark fields are suppressed. The current $\bar{u}\gamma_\alpha P_L d$ is conserved in the limit $m_{u,d} \to 0$, and consequently its strong interaction matrix elements are subtraction-point independent. Taking the $\pi^- \to \mu\bar{\nu}_\mu$ matrix element of Eq. (1.145) gives the pion decay amplitude

$$\mathcal{M} = -i\sqrt{2}G_F V_{ud} f_\pi\, \bar{u}(p_\mu)\not{p}_\pi P_L v(p_{\nu_\mu}), \tag{1.146}$$

where the pion decay constant, f_π, is the value of the pion-to-vacuum matrix element of the axial current,

$$\langle 0|\bar{u}\gamma^\alpha\gamma_5 d|\pi^-(p_\pi)\rangle = -if_\pi p_\pi^\alpha. \tag{1.147}$$

The measured pion decay rate gives $f_\pi \simeq 131$ MeV. In Eq. (1.147) the pion field is normalized by using the standard covariant norm: $\langle\pi(p'_\pi)|\pi(p_\pi)\rangle = 2E_\pi(2\pi)^3 \delta^3(\mathbf{p}'_\pi - \mathbf{p}_\pi)$. Parity invariance of the strong interactions implies that only the axial current part of the left-handed current contributes in Eq. (1.146).

In the limit $m_{u,d,s} = 0$, global $SU(3)_L$ transformations are a symmetry of QCD. The conserved currents associated with this symmetry can be derived by considering the change in the QCD Lagrangian under infinitesimal local $SU(3)_L$ transformations,

$$L = 1 + i\epsilon_L^A T^A, \tag{1.148}$$

with space–time dependent infinitesimal parameters $\epsilon_L^A(x)$. The change in the QCD Lagrange density, Eq. (1.93), under this transformation is

$$\delta\mathcal{L}_{\text{QCD}} = -J_{L\mu}^A \partial^\mu \epsilon_L^A, \tag{1.149}$$

where

$$J_{L\mu}^A = \bar{q}_L T^A \gamma_\mu q_L \tag{1.150}$$

are the conserved currents associated with $SU(3)_L$ transformations. We also know how left-handed transformations act on the meson fields in Σ. The change in the chiral Lagrange density under an infinitesimal left-handed transformation on the Σ in Eq. (1.98) is

$$\delta\mathcal{L}_{\text{eff}} = -J_{L\mu}^A \partial^\mu \epsilon_L^A, \tag{1.151}$$

where

$$J^A_{L\mu} = -\frac{if^2}{4}\mathrm{Tr}\, T^A\, \Sigma\, \partial_\mu \Sigma^\dagger. \tag{1.152}$$

Comparing Eqs. (1.150) and (1.152) gives

$$\bar{q}_L T^A \gamma_\mu q_L = -i\frac{f^2}{4}\mathrm{Tr}\, T^A \Sigma \partial_\mu \Sigma^\dagger + \cdots, \tag{1.153}$$

where the ellipses are contributions from higher derivative terms in the chiral Lagrangian. Matrix elements of the quark current involving the pseudo-Goldstone boson can be calculated by expanding Σ in terms of M on the right-hand side of Eq. (1.153). In particular, the part linear in M yields the tree-level relation $f = f_\pi$. Loops and higher derivatives operators in the chiral Lagrangian give corrections to the relation between f and f_π. The kaon decay constant is defined by

$$\langle 0|\bar{u}\gamma^\alpha\gamma_5 s|K^-(p_K)\rangle = -if_K p_K^\alpha. \tag{1.154}$$

The measured $K^- \to \mu\bar{\nu}_\mu$ decay rate determines f_K to be ~25% larger than the pion decay constant, $f_K \simeq 164$ MeV. At leading order in chiral $SU(3)_L \times SU(3)_R$, $f = f_K = f_\pi$, and the 25% difference between f_π and f_K is the typical size of $SU(3)_V$ breaking arising from the nonzero value of the strange quark mass.

At higher orders in chiral perturbation theory, the Noether procedure for finding the representation of $\bar{q}_L T^A \gamma_\mu q_L$ in terms of pseudo-Goldstone boson fields becomes ambiguous. Total derivative operators in the chiral Lagrangian can give a contribution to the current $J^A_{L\mu}$ (although not to the charges $Q^A_L = \int d^3x\, J^A_{L0}$), even though they are usually omitted from the chiral Lagrangian because they do not contribute to pseudo-Goldstone boson S-matrix elements. Note that at leading order in chiral perturbation theory there are no possible total derivative operators since $\partial^\mu(\mathrm{Tr}\, \Sigma^\dagger\partial_\mu\Sigma) = 0$.

1.8 The operator product expansion

The operator product expansion (OPE) is an important tool in particle physics and condensed matter physics, and it will be applied later in this book to describe inclusive B decay and to discuss sum rules. The use of the operator product expansion is best illustrated by an explicit example. In this section, the OPE will be applied to the study of deep inelastic lepton–proton scattering. The main purpose of the discussion is to explain the use of the OPE, so the presentation of the phenomenology of deep inelastic scattering will be kept to a minimum.

The basic deep inelastic scattering process is $\ell(k) + \mathrm{proton}(p) \to \ell(k') + X(p+q)$, in which an incoming lepton ℓ with momentum k scatters off a target proton, to produce an outgoing lepton ℓ with momentum k', plus anything X. The Feynman graph in Fig. 1.9 is the leading term in an expansion in the

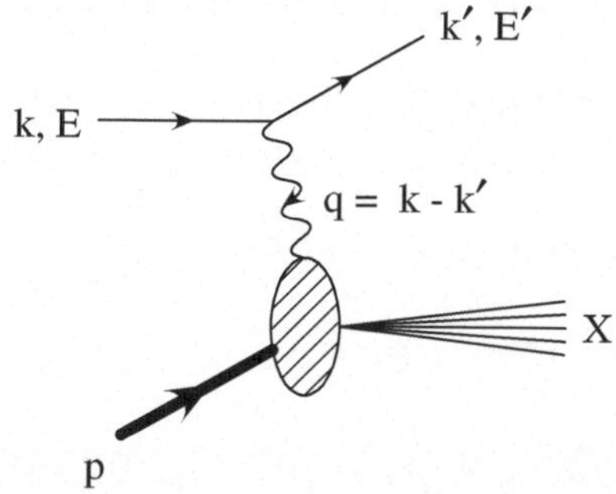

Fig. 1.9. The basic diagram for deep inelastic lepton–hadron scattering. The virtual photon momentum is q. The final hadronic state is not measured and is denoted by X.

electromagnetic fine structure constant α. The traditional kinematical variables used to describe the inclusive scattering process are the momentum transfer $Q^2 = -(k' - k)^2$, and the dimensionless variable x defined by

$$x = \frac{Q^2}{2p \cdot q}, \tag{1.155}$$

where $q = k - k'$. Note that for deep inelastic scattering, $Q^2 > 0$. It is also useful to define $\omega = 1/x$. The deep inelastic scattering cross section is the inclusive cross section in the limit that Q^2 is large with x fixed. The total cross section is obtained from squaring the amplitude represented by Fig. 1.9 and performing the appropriate phase space integrations. The lepton and photon parts of this amplitude as well as the phase space integrals can easily be computed. The nontrivial quantity is the square of the hadronic part of the diagram, which is

$$\sum_X (2\pi)^4 \delta^4(q + p - p_X) \langle p | J^\mu_{\text{em}}(0) | X \rangle \langle X | J^\nu_{\text{em}}(0) | p \rangle, \tag{1.156}$$

where the sum is over all possible final states X, and J^μ_{em} is the electromagnetic current. For convenience momentum and spin labels on the state vectors are suppressed. A spin average over the proton states $|p\rangle$ is also understood.

It is conventional to define the hadronic tensor

$$W^{\mu\nu}(p, q) = \frac{1}{4\pi} \int d^4x \, e^{iq \cdot x} \langle p | \left[J^\mu_{\text{em}}(x), J^\nu_{\text{em}}(0) \right] | p \rangle. \tag{1.157}$$

Inserting a complete set of states gives

$$W^{\mu\nu}(p, q) = \frac{1}{4\pi} \sum_X \int d^4x \, e^{iq \cdot x} \big[\langle p | J^\mu_{\text{em}}(x) | X \rangle \langle X | J^\nu_{\text{em}}(0) | p \rangle - \langle p | J^\nu_{\text{em}}(0) | X \rangle \langle X | J^\mu_{\text{em}}(x) | p \rangle \big], \tag{1.158}$$

where the sum on X is a sum over all final states, as well as an integral over the allowed final state phase space. Translation invariance implies that

$$\begin{aligned} \langle p|J^{\mu}_{\rm em}(x)|X\rangle &= \langle p|J^{\mu}_{\rm em}(0)|X\rangle e^{i(p-p_X)\cdot x}, \\ \langle X|J^{\mu}_{\rm em}(x)|p\rangle &= \langle X|J^{\mu}_{\rm em}(0)|p\rangle e^{i(p_X-p)\cdot x}. \end{aligned} \tag{1.159}$$

Inserting Eq. (1.159) into Eq. (1.158) gives

$$\begin{aligned} W^{\mu\nu}(p,q) = \frac{1}{4\pi}\sum_X \big[&(2\pi)^4\delta^4(q+p-p_X)\langle p|J^{\mu}_{\rm em}(0)|X\rangle\langle X|J^{\nu}_{\rm em}(0)|p\rangle \\ &-(2\pi)^4\delta^4(q+p_X-p)\langle p|J^{\nu}_{\rm em}(0)|X\rangle\langle X|J^{\mu}_{\rm em}(0)|p\rangle\big]. \end{aligned} \tag{1.160}$$

The only allowed final states are those with $p_X^0 \geq p^0$, since the baryon number is conserved. For $q^0 > 0$, only the first delta function in Eq. (1.160) can be satisfied, and the sum in $W_{\mu\nu}$ reduces to the expression in Eq. (1.156) involving the hadronic currents, and the energy-momentum conserving delta function (up to a factor of $1/4\pi$). Since only the first term in Eq. (1.158) contributes, one could have defined $W_{\mu\nu}$ in Eq. (1.157) simply as the matrix element of $J^{\mu}_{\rm em}(x)J^{\nu}_{\rm em}(0)$ without the commutator. The reason for using the commutator is that then $W_{\mu\nu}$ has a nicer analytic structure when continued away from the physical region. The most general form of $W_{\mu\nu}$ consistent with current conservation, parity and time-reversal invariance is

$$W_{\mu\nu} = F_1\left(-g_{\mu\nu} + \frac{q_\mu q_\nu}{q^2}\right) + \frac{F_2}{p\cdot q}\left(p_\mu - \frac{p\cdot q\, q_\mu}{q^2}\right)\left(p_\nu - \frac{p\cdot q\, q_\nu}{q^2}\right), \tag{1.161}$$

where $F_{1,2}$ can be written as functions of x and Q^2. Here $F_{1,2}$ are called structure functions.

The Q^2 dependence of the structure functions can be calculated in quantum chromodynamics. The starting point in the derivation is the time-ordered product of two currents:

$$t^{\mu\nu} \equiv i\int d^4x e^{iq\cdot x}\, T\left[J^{\mu}_{\rm em}(x)\, J^{\nu}_{\rm em}(0)\right]. \tag{1.162}$$

The proton matrix element of $t_{\mu\nu}$,

$$T_{\mu\nu} = \langle p|t_{\mu\nu}|p\rangle, \tag{1.163}$$

can also be written in terms of structure functions,

$$T_{\mu\nu} = T_1\left(-g_{\mu\nu} + \frac{q_\mu q_\nu}{q^2}\right) + \frac{T_2}{p\cdot q}\left(p_\mu - \frac{p\cdot q\, q_\mu}{q^2}\right)\left(p_\nu - \frac{p\cdot q\, q_\nu}{q^2}\right). \tag{1.164}$$

The analytic structure of $T_{1,2}$ as a function of ω for fixed Q^2 is shown in Fig. 1.10. There are cuts in the physical region $1 \leq |\omega|$. The discontinuity across the

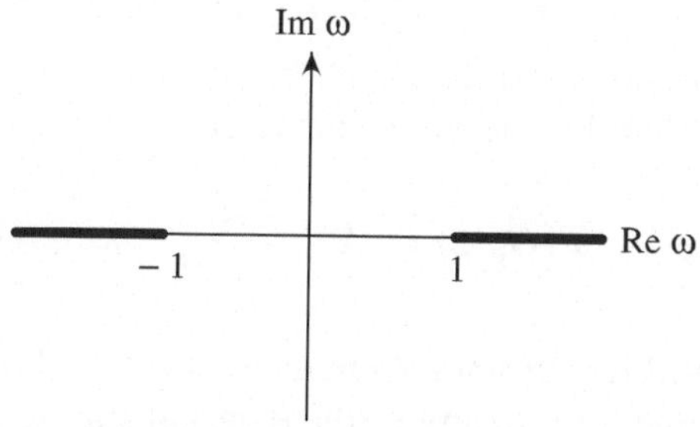

Fig. 1.10. The analytic structure of $T_{\mu\nu}$ in the complex ω plane. The discontinuity across the cuts $1 \leq |\omega| \leq \infty$ is related to $W_{\mu\nu}$.

right-hand cut for $T_{1,2}$ is $F_{1,2}$,

$$\text{Im}\, T_{1,2}(\omega + i\epsilon, Q^2) = 2\pi F_{1,2}(\omega, Q^2). \tag{1.165}$$

[The discontinuity across the left-hand cut gives the structure functions for deep inelastic scattering off antiprotons.]

The key idea that permits the computation of $T_{\mu\nu}$ in certain limiting cases is the operator product expansion. Consider the time-ordered product of two local operators separated in position by z:

$$T\,[O_a(z)\; O_b(0)]\,. \tag{1.166}$$

For small z, the operators are at practically the same point. In this limit, the operator product can be written as an expansion in local operators,

$$T\,[O_a(z)\; O_b(0)] = \sum_k C_{abk}(z) O_k(0). \tag{1.167}$$

The coefficient functions depend on the separation z. Low-momentum (compared with $1/z$) matrix elements of the left-hand side are completely equivalent to matrix elements of the right-hand side. Thus one can replace the product $T\,[O_a(z) O_b(0)]$ in the computation of matrix elements by the expansion in Eq. (1.167), where the coefficients $C_{abk}(z)$ are independent of the matrix elements, provided that the external states have momentum components that are small compared with the inverse separation $1/z$. In QCD, the coupling constant is small at short distances because of asymptotic freedom. Thus the coefficient functions can be computed in perturbation theory, since all nonperturbative effects occur at scales that are much larger than z, and do not affect the computation of the coefficient functions.

The momentum space version of the operator product expansion is for the product

$$\int d^4 z\, e^{iq\cdot z} T[O_a(z)\; O_b(0)]. \tag{1.168}$$

In the limit that $q \to \infty$, the Fourier transform in Eq. (1.168) forces $z \to 0$, and again the operator product can be expanded in terms of local operators with coefficient functions that depend on q. For large q,

$$\int d^4 z e^{iq\cdot z} T[O_a(z)\, O_b(0)] = \sum_k C_{abk}(q) O_k(0). \tag{1.169}$$

This expansion is valid for all matrix elements, provided q is much larger than the characteristic momentum in any of the external states.

We will use the Fourier transform version of the operator product expansion, Eq. (1.169). The product of two electromagnetic currents in Eq. (1.162) can be expanded in terms of a sum of local operators multiplied by coefficients that are functions of q. This expansion will be valid for proton matrix elements, Eq. (1.163), provided that q is much larger than the typical hadronic mass scale $\Lambda_{\rm QCD}$. The local operators in the operator product expansion for QCD are quark and gluon operators with arbitrary dimension d and spin n. An operator with spin n and dimension d can be written as $O_{d,n}^{\mu_1\cdots\mu_n}$, where $O_{d,n}$ is symmetric and traceless in $\mu_1 \cdots \mu_n$. The matrix element of $O_{d,n}$ in the spin-averaged proton target is proportional to $m_p^{d-n-2}\, \mathcal{S}[p^{\mu_1} \cdots p^{\mu_n}]$. $\mathcal{S}$ acts on a tensor to project out the completely symmetric traceless component. The power of m_p follows from dimensional analysis, since a proton state with the conventional relativistic normalization has dimension minus one. The coefficient functions in the operator product expansion are functions only of q. Thus the free indices on the operator O must be either μ, ν or be contracted with q^α. Every index on O contracted with q^α produces a factor of $p \cdot q$, which is of the order of Q^2 in the deep inelastic limit. An index μ or ν is contracted with the lepton momentum, and produces a factor of $p \cdot k$ or $p \cdot k'$, both of which are also of the order of Q^2 in the deep inelastic limit. In addition, since $t_{\mu\nu}$ has dimension two, the coefficient of O must have dimension $[\text{mass}]^{2-d}$ in the operator product expansion. Thus, the contribution of any operator O to the differential cross section is of the order of

$$\begin{aligned}
C_{\mu_1\cdots\mu_n} O_{d,n}^{\mu_1\cdots\mu_n} &\to \frac{q_{\mu_1}}{Q} \cdots \frac{q_{\mu_n}}{Q} Q^{2-d} \langle O_{d,n}^{\mu_1\cdots\mu_n} \rangle, \\
&\to \frac{q_{\mu_1}}{Q} \cdots \frac{q_{\mu_n}}{Q} Q^{2-d} m_p^{d-n-2}\, p^{\mu_1} \cdots p^{\mu_n}, \\
&\to \frac{(p\cdot q)^n}{Q^n} Q^{2-d} m_p^{d-n-2}, \\
&\to \omega^n \left(\frac{Q}{m_p}\right)^{2+n-d} = \omega^n \left(\frac{Q}{m_p}\right)^{2-t},
\end{aligned} \tag{1.170}$$

where the twist t is defined as

$$t = d - n = \text{dimension} - \text{spin}. \tag{1.171}$$

Table 1.2. *Dimension, spin, and twist for the basic objects in the QCD Lagrangian*

Parameter	q	$G_{\mu\nu}$	D^μ
Dimension	3/2	2	1
Spin	1/2	1	1
Twist	1	1	0

The most important operators in the operator product expansion are those with the lowest possible twist. Twist-two operators contribute a finite amount to the structure functions in the deep inelastic limit, twist three contributions are suppressed by m_p/Q, and so on. The fundamental fields in QCD are quark and gluon fields, so the gauge invariant operators in the operator product expansion can be written in terms of quark fields q, the gluon field strength $G_{\mu\nu}$, and the covariant derivative D^μ. Table 1.2 lists the basic objects, with their dimension and twist. Any gauge invariant operator must contain at least two quark fields, or two gluon field strength tensors. Thus the lowest possible twist is two. A twist-two operator has either two q's or two $G_{\mu\nu}$'s and an arbitrary number of covariant derivatives. The indices of the covariant derivatives are not contracted, because an operator such as D^2 has twist two, whereas the traceless symmetric part of $D^\alpha D^\beta$ has twist zero.

The first step in doing an operator product expansion is to determine all the linearly independent operators that can occur. We have just seen that the leading operators are twist-two quark and gluon operators. We will simplify the analysis by considering not the electromagnetic current but rather $J_\mu = \bar{q}\gamma_\mu q$ for a single quark flavor q. Results for the realistic case can be obtained by summing over flavors weighted by the square of quark charges. The Lorentz structure of the quark operators must be either $\bar{q}\gamma^\mu q$ or $\bar{q}\gamma^\mu\gamma_5 q$ in the limit that light quark masses can be neglected, because the operator product $J^\mu J^\nu$ does not change chirality. The conventional basis for twist-two quark operators is:

$$O_{q,V}^{\mu_1\cdots\mu_n} = \frac{1}{2}\left(\frac{i}{2}\right)^{n-1} \mathcal{S}\left\{\bar{q}\,\gamma^{\mu_1} \overleftrightarrow{D}^{\mu_2} \cdots \overleftrightarrow{D}^{\mu_n} q\right\}, \tag{1.172}$$

$$O_{q,A}^{\mu_1\cdots\mu_n} = \frac{1}{2}\left(\frac{i}{2}\right)^{n-1} \mathcal{S}\left\{\bar{q}\,\gamma^{\mu_1} \overleftrightarrow{D}^{\mu_2} \cdots \overleftrightarrow{D}^{\mu_n} \gamma_5 q\right\}, \tag{1.173}$$

where

$$\bar{A} \overleftrightarrow{D}^{\mu} B = \bar{A}\overrightarrow{D}^\mu B - \bar{A}\overleftarrow{D}^\mu B. \tag{1.174}$$

The operators $O_{q,A}^{\mu_1\cdots\mu_n}$ have matrix elements proportional to the proton spin, and so do not contribute to spin-averaged scattering. The tower of twist-two gluon operators needed for scattering from unpolarized protons is

$$O_{g,V}^{\mu_1\cdots\mu_n} = -\frac{1}{2}\left(\frac{i}{2}\right)^{n-2} \mathcal{S}\left\{G_A^{\mu_1\alpha} \overleftrightarrow{D}^{\mu_2} \ldots \overleftrightarrow{D}^{\mu_{n-1}} G_{A\alpha}{}^{\mu_n}\right\}. \tag{1.175}$$

We will only compute the operator product expansion to lowest order in α_s, so the gluon operators do not occur.

The most general form for $t^{\mu\nu}$ consistent with current conservation and using only twist-two operators is

$$\begin{aligned} t_{\mu\nu} = & \sum_{n=2,4,\ldots}^{\infty} \left(-g_{\mu\nu} + \frac{q_\mu q_\nu}{q^2}\right) \frac{2^n q_{\mu_1}\cdots q_{\mu_n}}{(-q^2)^n} \sum_{j=q,g} 2\, C_{j,n}^{(1)}\, O_{j,V}^{\mu_1\cdots\mu_n} \\ & + \sum_{n=2,4,\ldots}^{\infty} \left(g_{\mu\mu_1} - \frac{q_\mu q_{\mu_1}}{q^2}\right)\left(g_{\nu\mu_2} - \frac{q_\nu q_{\mu_2}}{q^2}\right) \\ & \times \frac{2^n q_{\mu_3}\cdots q_{\mu_n}}{(-q^2)^{n-1}} \sum_{j=q,g} 2C_{j,n}^{(2)} O_{j,V}^{\mu_1\cdots\mu_n}, \end{aligned} \tag{1.176}$$

where the unknown coefficients are $C_{j,n}^{(1)}$ and $C_{j,n}^{(2)}$, and the factors of two and signs have been chosen for later convenience.

The second step in doing an operator product expansion is to determine the coefficients of the operators, $C_{j,n}^{(1)}$ and $C_{j,n}^{(2)}$. The best way to do this is to evaluate enough on-shell matrix elements to determine all the coefficients. Since we have argued that the coefficients can be computed using any matrix elements, we will evaluate the coefficients by taking matrix elements in on-shell quark and gluon states. We will only illustrate the computation of the coefficients to lowest nontrivial order, i.e., $(\alpha_s)^0$, in this chapter.

A generic term in the operator product expansion can be written as

$$JJ \sim C_q O_q + C_g O_g, \tag{1.177}$$

where q and g refer to quark and gluon operators. Taking the matrix element of both sides in a free quark state gives

$$\langle q|JJ|q\rangle \sim C_q\langle q|O_q|q\rangle + C_g\langle q|O_g|q\rangle. \tag{1.178}$$

The electromagnetic current is a quark operator. Thus the left-hand side is of the order of $(\alpha_s)^0$. The matrix element $\langle q|O_q|q\rangle$ is also of the order of $(\alpha_s)^0$, whereas the matrix element $\langle q|O_g|q\rangle$ is of the order of $(\alpha_s)^1$ since there are at least two gluons in O_g, each of which contributes a factor of the QCD coupling constant g to the matrix element. Thus, one can determine C_q to leading order by taking the quark matrix element of both sides of the operator product expansion, neglecting the gluon operators.

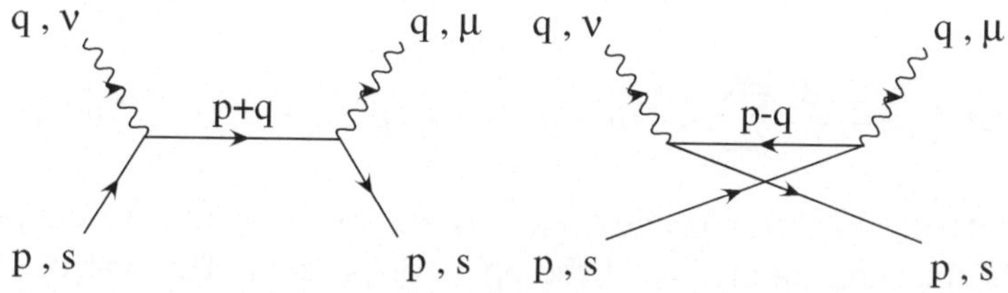

Fig. 1.11. The lowest order diagrams contributing to the quark matrix element of the product of two electromagnetic currents.

As mentioned previously, we work in a theory with a single quark flavor with charge one. The quark matrix element of the left-hand side of the operator product expansion, Eq. (1.169), is given by the Feynman graphs in Fig. 1.11,

$$\mathcal{M}^{\mu\nu} = i\bar{u}(p,s)\,\gamma^\mu i\frac{\not p + \not q}{(p+q)^2}\gamma^\nu u(p,s) + i\bar{u}(p,s)\,\gamma^\nu i\frac{\not p - \not q}{(p-q)^2}\gamma^\mu u(p,s). \tag{1.179}$$

Note that there is an overall factor of i because we are computing i times the time-ordered product in Eq. (1.162). The crossed diagram (second term) can be obtained by the replacement $\mu \leftrightarrow \nu, q \to -q$ from the direct diagram (first term), so we concentrate on simplifying the first term. Expanding the denominator gives

$$(p+q)^2 = 2p\cdot q + q^2 = q^2\left(1 + \frac{2p\cdot q}{q^2}\right) = q^2(1-\omega)\,, \tag{1.180}$$

since $p^2 = 0$ for an on-shell massless quark. The numerator can be simplified using the γ matrix identity in Eq. (1.119):

$$\begin{aligned}\bar{u}(p,s)\gamma^\mu(\not p + \not q)\gamma^\nu u(p,s) = \bar{u}(p,s)[&(p+q)^\mu\gamma^\nu + (p+q)^\nu\gamma^\mu \\ &- g^{\mu\nu}(\not p + \not q) + i\epsilon^{\mu\nu\alpha\lambda}(p+q)_\alpha\gamma_\lambda\gamma_5]u(p,s).\end{aligned} \tag{1.181}$$

For an on-shell massless quark,

$$\not p\, u(p,s) = 0, \quad \bar{u}(p,s)\,\gamma_\lambda\, u(p,s) = 2p_\lambda, \quad \bar{u}(p,s)\,\gamma_\lambda\gamma_5\, u(p,s) = 2h\; p_\lambda, \tag{1.182}$$

where h is the quark helicity. Thus the $\not p$ and $\epsilon^{\mu\nu\alpha\lambda}\, p_\alpha\gamma_\lambda\gamma_5$ terms both give zero. For spin-averaged matrix elements the sum over helicities gives zero and so we neglect the part of $\mathcal{M}^{\mu\nu}$ proportional to h. Combining the various terms and using

$$(1-\omega)^{-1} = \sum_{n=0}^{\infty}\omega^n \tag{1.183}$$

gives

$$\mathcal{M}^{\mu\nu} = -\frac{2}{q^2}\sum_{n=0}^{\infty} \omega^n [(p+q)^\mu p^\nu + (p+q)^\nu p^\mu - g^{\mu\nu} p \cdot q]. \quad (1.184)$$

To complete the operator product expansion, we need the free quark matrix element of the right-hand side of the operator product. The matrix element of the quark operators of Eq. (1.172) in a free quark state with momentum p is

$$\langle q\,(p)| O_{q,V}^{\mu_1\cdots\mu_n} |q\,(p)\rangle = \mathcal{S}[p^{\mu_1}\cdots p^{\mu_n}] = p^{\mu_1}\cdots p^{\mu_n}, \quad (1.185)$$

since $p^2 = 0$. The factors of i and 2 in Eqs. (1.172) and (1.173) were chosen so that no such factors appear in the matrix elements.

We determine the coefficient functions for the spin-independent terms in the operator product expansion. Including the crossed diagram, the spin-independent terms on the left-hand side of the operator product are

$$\begin{aligned}\mathcal{M}^{\mu\nu} &= -\frac{2}{q^2}\sum_{n=0}^{\infty} \omega^n [(p+q)^\mu p^\nu + (p+q)^\nu p^\mu - g^{\mu\nu} p \cdot q] \\ &\quad + (\mu \leftrightarrow \nu,\ q \to -q,\ \omega \to -\omega), \end{aligned} \quad (1.186)$$

since ω is odd in q. The crossed diagram causes half the terms to cancel, so that the matrix element is

$$\begin{aligned}\mathcal{M}^{\mu\nu} &= -\frac{4}{q^2}\sum_{n=0,2,4}^{\infty} \omega^n 2 p^\mu p^\nu - \frac{4}{q^2}\sum_{n=1,3,5}^{\infty} \omega^n (q^\mu p^\nu + q^\nu p^\mu - g^{\mu\nu} p \cdot q) \\ &= -\frac{8}{q^2}\sum_{n=0,2,4}^{\infty} \frac{2^n (p\cdot q)^n}{(-q^2)^n}\left(p^\mu - \frac{p\cdot q q^\mu}{q^2}\right)\left(p^\nu - \frac{p\cdot q q^\nu}{q^2}\right) \\ &\quad - \frac{4}{q^2}\sum_{n=1,3,5}^{\infty} \frac{2^n (p\cdot q)^{n+1}}{(-q^2)^n}\left(-g^{\mu\nu} + \frac{q^\mu q^\nu}{q^2}\right). \end{aligned} \quad (1.187)$$

Equation (1.187) can be rewritten in the form

$$\begin{aligned}\mathcal{M}^{\mu\nu} &= -\frac{8}{q^2}\sum_{n=0,2,4}^{\infty} \frac{2^n q^{\mu_3}\cdots q^{\mu_{n+2}}}{(-q^2)^n} \\ &\quad \times \left(g^{\mu\mu_1} - \frac{q^\mu q^{\mu_1}}{q^2}\right)\left(g^{\nu\mu_2} - \frac{q^\nu q^{\mu_2}}{q^2}\right) p_{\mu_1}\cdots p_{\mu_{n+2}} \\ &\quad - \frac{4}{q^2}\sum_{n=1,3,5}^{\infty} \frac{2^n q^{\mu_1}\cdots q^{\mu_{n+1}}}{(-q^2)^n}\left(-g^{\mu\nu} + \frac{q^\mu q^\nu}{q^2}\right) p_{\mu_1}\cdots p_{\mu_{n+1}}, \end{aligned} \quad (1.188)$$

which separates the q and p dependence.

The coefficient functions in the operator product depend only on q, and the matrix elements depend only on p. We have separated the operator product into pieces which depend only on q and only on p. By comparing with Eq. (1.185), we can write Eq. (1.188) as

$$\begin{aligned}\mathcal{M}^{\mu\nu} = &-\frac{8}{q^2}\sum_{n=0,2,4}^{\infty}\frac{2^n q^{\mu_3}\cdots q^{\mu_{n+2}}}{(-q^2)^n} \\ &\times\left(g^{\mu\mu_1}-\frac{q^\mu q^{\mu_1}}{q^2}\right)\left(g^{\nu\mu_2}-\frac{q^\nu q^{\mu_2}}{q^2}\right)\langle p|O_{q,V\mu_1\cdots\mu_{n+2}}|p\rangle \\ &-\frac{4}{q^2}\sum_{n=1,3,5}^{\infty}\frac{2^n q^{\mu_1}\cdots q^{\mu_{n+1}}}{(-q^2)^n}\left(-g^{\mu\nu}+\frac{q^\mu q^\nu}{q^2}\right)\langle p|O_{q,V\mu_1\cdots\mu_{n+1}}|p\rangle\end{aligned} \tag{1.189}$$

so that

$$\begin{aligned}t^{\mu\nu} = &\,2\sum_{n=2,4,6}^{\infty}\frac{2^n q^{\mu_3}\cdots q^{\mu_n}}{(-q^2)^{n-1}}\left(g^{\mu\mu_1}-\frac{q^\mu q^{\mu_1}}{q^2}\right)\left(g^{\nu\mu_2}-\frac{q^\nu q^{\mu_2}}{q^2}\right)O_{q,V\mu_1\cdots\mu_n} \\ &+2\sum_{n=2,4,6}^{\infty}\frac{2^n q^{\mu_1}\cdots q^{\mu_n}}{(-q^2)^n}\left(-g^{\mu\nu}+\frac{q^\mu q^\nu}{q^2}\right)O_{q,V\mu_1\cdots\mu_n}.\end{aligned} \tag{1.190}$$

This is the operator product expansion for the spin-independent part of $t_{\mu\nu}$, i.e., the part involving only vector operators. Only vector operators with n even occur in the operator product expansion, because $t^{\mu\nu}$ is even under charge conjugation.

Comparing with the most general form for the operator product in Eq. (1.176), we see that at lowest order in α_s the coefficients $C_{q,n}^{(1,2)} = 1$. Considering a gluon matrix element gives $C_{g,n} = 0$ at lowest order in α_s. At higher orders in α_s the coefficient functions and the operator matrix elements depend on a subtraction point μ. Since the physical quantity $t_{\mu\nu}$ is independent of the arbitrary choice of subtraction point, a renormalization group equation similar to that for coefficients in the weak nonleptonic decay Hamiltonian in Eq. (1.132) can be derived for the coefficients $C_{j,n}^{(1,2)}$. At $\mu = Q$ there are no large logarithms in the coefficients $C_{j,n}^{(1,2)}$. Therefore, we have that

$$\begin{aligned}C_{q,n}^{(1,2)}[1, \alpha_s(Q)] &= 1 + \mathcal{O}[\alpha_s(Q)], \\ C_{g,n}^{(1,2)}[1, \alpha_s(Q)] &= 0 + \mathcal{O}[\alpha_s(Q)].\end{aligned} \tag{1.191}$$

However, at $\mu = Q$ there are large logarithms of Q/Λ_{QCD} in the nucleon matrix element of the twist-two operators. It is convenient to use the renormalization group equations that the $C_{j,n}^{(1,2)}$ satisfy and the initial conditions in Eqs. (1.191) to move the Q dependence from the matrix elements into the coefficients by scaling the subtraction point down to a value $\mu \ll Q$. It is this calculable Q dependence that results in the dependence of the structure functions $T_{1,2}$ and hence $F_{1,2}$ on Q, without which they would just be functions of x. So quantum chromodynamics predicts a calculable logarithmic dependence of the structure functions $F_{1,2}$ on Q,

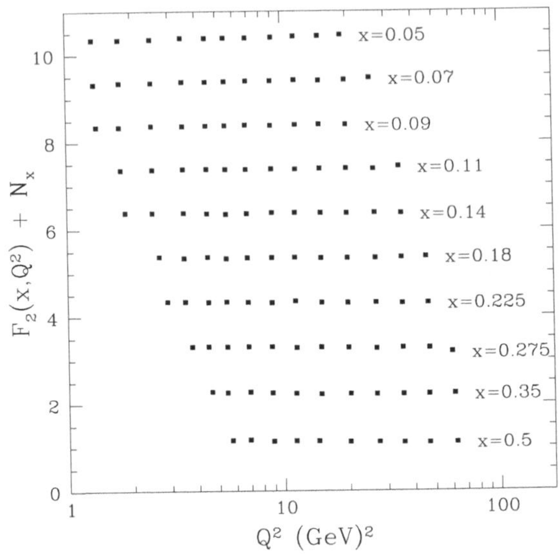

Fig. 1.12. The proton structure function $F_2(x, Q^2)$, measured in deep inelastic muon scattering by the NMC Collaboration [M. Arneodo et al., Phys. Lett. 364B (1995) 107]. The data is shown as a function of Q^2 for different values of x. For clarity, the plots for different x values are offset by one unit vertically, so that what is plotted is $F_2 + N_x$, where N_x is an integer equal to 1 for $x = 0.5$, 2 for $x = 0.35$, etc.

which has been verified experimentally. The fact that this dependence is weak at large Q is a consequence of asymptotic freedom. In free field theory the structure functions $F_{1,2}$ are independent of Q, which is called scaling. The logarithmic Q dependence is usually called a scaling violation. Some experimental data showing the approximate scaling of F_2 are shown in Fig. 1.12.

1.9 Problems

1. Consider an $SU(5)$ gauge theory with a scalar field Φ that transforms in the adjoint representation

$$\Phi \to U\Phi U^\dagger, \quad U \in SU(5).$$

Suppose Φ gets the vacuum expectation value

$$\langle\Phi\rangle = v \begin{bmatrix} 2 & 0 & 0 & 0 & 0 \\ 0 & 2 & 0 & 0 & 0 \\ 0 & 0 & 2 & 0 & 0 \\ 0 & 0 & 0 & -3 & 0 \\ 0 & 0 & 0 & 0 & -3 \end{bmatrix}.$$

(a) What is the unbroken subgroup H of $SU(5)$?
(b) What are the H quantum numbers of the massive $SU(5)$ gauge bosons?

2. If there are N generations of quarks and leptons, show that the CKM matrix contains $(N-1)^2$ real parameters.
3. Calculate the vertex renormalization constant Z_e given in Eq. (1.62).
4. Calculate to order g^2 the renormalization matrix Z_{ij} defined in Eq. (1.127) for the operators O_1 and O_2 defined in Eq. (1.125). Use it to deduce the anomalous dimension matrix in Eq. (1.135).
5. Calculate the cross section $\sigma(\pi^+\pi^- \to \pi^+\pi^-)$ at center of mass energy E to leading order in the chiral perturbation theory expansion.
6. In chiral perturbation theory, any Feynman diagram contributing to π–π scattering has L loops, n_k insertions of vertices of order p^k, and N_π internal pion lines. The resulting amplitude is of order p^D, where

$$D = (\text{powers of } p \text{ in numerator}) - (\text{powers of } p \text{ in denominator}).$$

Using the identity $L = N_\pi - \sum_k n_k + 1$, derive Eq. (1.110) for D.

7. Calculate the decay amplitude for $K^- \to \pi^0 e\bar{\nu}_e$ at leading order in chiral perturbation theory.
8. (a) Calculate the semileptonic free quark decay rate $\Gamma(b \to ce\bar{\nu}_e)$.
   (b) Using the renormalization group improved effective Hamiltonian in Eq. (1.124), calculate the nonleptonic free quark decay rate $\Gamma(b \to cd\bar{u})$.

   Neglect all masses except those of the b and c quarks.

1.10 References

The material in this chapter can be found in many textbooks. Some suggested references are:

J.D. Bjorken and S.D. Drell, *Relativistic Quantum Mechanics*, McGraw-Hill, 1964
J.D. Bjorken and S.D. Drell, *Relativistic Quantum Fields*, McGraw-Hill, 1964
C. Itzykson and J.-B. Zuber, *Quantum Field Theory*, McGraw-Hill, 1980
M.E. Peskin and D.V. Schroeder, *An Introduction to Quantum Field Theory*, Addison-Wesley, 1995
S. Weinberg, *The Quantum Theory of Fields, Vol I: Foundations and Vol II: Modern Applications*, Cambridge University Press, 1995
R. Balian and J. Zinn-Justin (editors), *Methods in Field Theory*, Les Houches Session XXVIII, North-Holland, 1976
H. Georgi, *Weak Interactions and Modern Particle Theory*, Benjamin/Cummings, 1984
J.F. Donoghue, E. Golowich, and B.R. Holstein, *Dynamics of the Standard Model*, Cambridge University Press, 1992

2
Heavy quarks

The light u, d, and s quarks have masses m_q that are small compared to the scale of nonperturbative strong dynamics. Consequently, it is a good approximation to take the $m_q \to 0$ limit of QCD. In this limit QCD has an $SU(3)_L \times SU(3)_R$ chiral symmetry, which can be used to predict some properties of hadrons containing these light quarks. For quarks with masses m_Q that are large compared with the scale of nonperturbative strong dynamics, it is a good approximation to take the $m_Q \to \infty$ limit of QCD. In this limit QCD has spin-flavor heavy quark symmetry, which has important implications for the properties of hadrons containing a single heavy quark.

2.1 Introduction

The QCD Lagrangian in Eq. (1.82) describes the strong interactions of light quarks and gluons. As discussed in Sec. 1.4, there is a nonperturbative scale Λ_{QCD} that is dynamically generated by QCD. A color singlet state, such as a meson made up of a quark–antiquark pair, is bound by the nonperturbative gluon dynamics. If the quarks are light, the typical size of such a system is of the order of $\Lambda_{\text{QCD}}^{-1}$. Consider a $Q\bar{q}$ meson that contains a heavy quark with mass $m_Q \gg \Lambda_{\text{QCD}}$, and a light quark with mass $m_q \ll \Lambda_{\text{QCD}}$. Such a heavy-light meson also has a typical size of the order of $\Lambda_{\text{QCD}}^{-1}$, as for mesons containing only light quarks. The typical momentum transfer between the heavy and light quarks in the $Q\bar{q}$ meson arising from nonperturbative QCD dynamics is of the order of Λ_{QCD}. An important consequence of this fact is that the velocity v of the heavy quark is almost unchanged by such strong interaction effects, even though the momentum p of the heavy quark changes by an amount of the order of Λ_{QCD}, since $\Delta v = \Delta p / m_Q$. A similar argument holds for any hadron containing a single heavy quark Q.

In the limit $m_Q \to \infty$, the heavy quark in the meson can be labeled by a velocity four-vector v that does not change with time. The heavy quark behaves like a static external source that transforms as a color triplet, and the meson dynamics reduces to that of light degrees of freedom interacting with this color source. One sees immediately that the mass of the heavy quark is completely irrelevant in the limit $m_Q \to \infty$, so that all heavy quarks interact in the same way within heavy mesons. This leads to *heavy quark flavor symmetry:* the dynamics is unchanged under the exchange of heavy quark flavors. The $1/m_Q$ corrections take into account finite mass effects and differ for quarks of different masses. As a result, heavy quark flavor symmetry breaking effects are proportional to $(1/m_{Q_i} - 1/m_{Q_j})$, where Q_i and Q_j are any two heavy quark flavors. The only strong interaction of a heavy quark is with gluons, as there are no quark–quark interactions in the Lagrangian. In the $m_Q \to \infty$ limit, the static heavy quark can only interact with gluons via its chromoelectric charge. This interaction is spin independent. This leads to *heavy quark spin symmetry:* the dynamics is unchanged under arbitrary transformations on the spin of the heavy quark. The spin-dependent interactions are proportional to the chromomagnetic moment of the quark, and so are of the order of $1/m_Q$. Heavy quark spin symmetry breaking does not have to be proportional to the difference of $1/m_Q$'s, since the spin symmetry is broken even if there are two heavy quarks with the same mass. The heavy quark $SU(2)$ spin symmetry and $U(N_h)$ flavor symmetries (for N_h heavy flavors) can be embedded into a larger $U(2N_h)$ spin-flavor symmetry in the $m_Q \to \infty$ limit. Under this symmetry the $2N_h$ states of the N_h heavy quarks with spin up and down transform as the fundamental representation. We will see in Sec. 2.6 that the effective Lagrangian can be written in a way that makes this symmetry manifest.

2.2 Quantum numbers

Heavy hadrons contain a heavy quark as well as light quarks and/or antiquarks and gluons. All the degrees of freedom other than the heavy quark are referred to as the light degrees of freedom ℓ. For example, a heavy $Q\bar{q}$ meson has an antiquark $\bar{q}$, gluons, and an arbitrary number of $\bar{q}q$ pairs as the light degrees of freedom. Although the light degrees of freedom are some complicated mixture of the antiquark $\bar{q}$, gluons, and $\bar{q}q$ pairs, they must have the quantum numbers of a single antiquark $\bar{q}$. The total angular momentum of the hadron $\mathbf{J}$ is a conserved operator. We have also seen that the spin of the heavy quark $\mathbf{S}_Q$ is conserved in the $m_Q \to \infty$ limit. Therefore, the spin of the light degrees of freedom $\mathbf{S}_\ell$ defined by

$$\mathbf{S}_\ell \equiv \mathbf{J} - \mathbf{S}_Q \tag{2.1}$$

is also conserved in the heavy quark limit. The light degrees of freedom in a hadron are quite complicated and include superpositions of states with different particle numbers. Nevertheless, the total spin of the light degrees of freedom is a good quantum number in heavy hadrons. We will define the quantum numbers j,

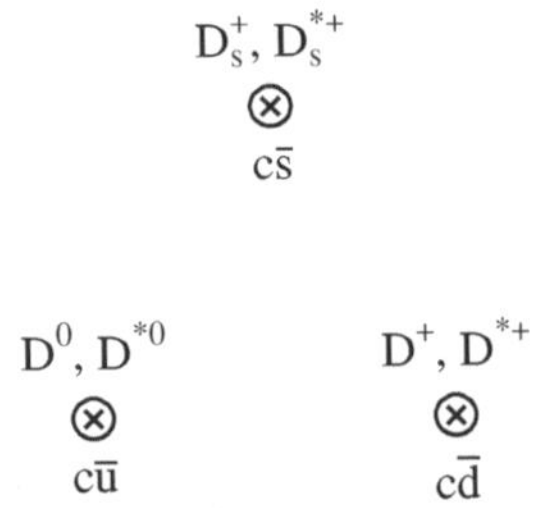

Fig. 2.1. Flavor $SU(3)$ weight diagram for the spin-0 pseudoscalar and spin-1 vector $c\bar{q}$ mesons. The corresponding $b\bar{q}$ mesons are the $\bar{B}_s^0$, $\bar{B}^-$, and $\bar{B}^0$, and their spin-1 partners. The vertical direction is hypercharge, and the horizontal direction is I_3, the third component of isospin.

s_Q, and s_ℓ as the eigenvalues $\mathbf{J}^2 = j\,(j+1)$, $\mathbf{S}_Q^2 = s_Q(s_Q+1)$, and $\mathbf{S}_\ell^2 = s_\ell\,(s_\ell+1)$ in the state $H^{(Q)}$. Heavy hadrons come in doublets (unless $s_\ell = 0$) containing states with total spin $j_\pm = s_\ell \pm 1/2$ obtained by combining the spin of the light degrees of freedom with the spin of the heavy quark $s_Q = 1/2$. These doublets are degenerate in the $m_Q \to \infty$ limit. If $s_\ell = 0$, there is only a single $j = 1/2$ hadron.

Mesons containing a heavy quark Q are made up of a heavy quark and a light antiquark $\bar{q}$ (plus gluons and $q\bar{q}$ pairs). The ground state mesons are composed of a heavy quark with $s_Q = 1/2$ and light degrees of freedom with $s_\ell = 1/2$, forming a multiplet of hadrons with spin $j = 1/2 \otimes 1/2 = 0 \oplus 1$ and negative parity, since quarks and antiquarks have opposite intrinsic parity. These states are the D and D^* mesons if Q is a charm quark, and the $\bar{B}$ and $\bar{B}^*$ mesons if Q is a b quark. The field operators which annihilate these heavy quark mesons with velocity v are denoted by $P_v^{(Q)}$ and $P_{v\mu}^{*(Q)}$, respectively. The light antiquark can be either a $\bar{u}$, $\bar{d}$, or $\bar{s}$ quark, so each of these heavy meson fields form a $\bar{\mathbf{3}}$ representation of the light quark flavor group $SU(3)_V$. The $SU(3)$ weight diagram for the $\bar{\mathbf{3}}$ mesons is shown in Fig. 2.1.

In the nonrelativistic constituent quark model, the first excited heavy meson states have a unit of orbital angular momentum between the constituent antiquark and the heavy quark. These $L = 1$ mesons have $s_\ell = 1/2$ or $3/2$, depending on how the orbital angular momentum is combined with the antiquark spin. The $s_\ell = 1/2$ mesons form multiplets of spin parity 0^+ and 1^+ states named (for $Q = c$) D_0^* and D_1^*, and the $s_\ell = 3/2$ mesons form multiplets of 1^+ and 2^+ states named (for $Q = c$) D_1 and D_2^*. Properties of the $s_\ell = 1/2$ and $s_\ell = 3/2$ states are related in the nonrelativistic constituent quark model, but not by heavy quark symmetry.

Baryons containing a heavy quark consist of a heavy quark and two light quarks, plus gluons and $q\bar{q}$ pairs. The lowest-lying baryons have $s_\ell = 0$ and $s_\ell = 1$ and form $\bar{\mathbf{3}}$ and $\mathbf{6}$ representations of $SU(3)_V$, which are shown in Figs. 2.2 and 2.3, respectively. We can easily understand this pattern in the nonrelativistic

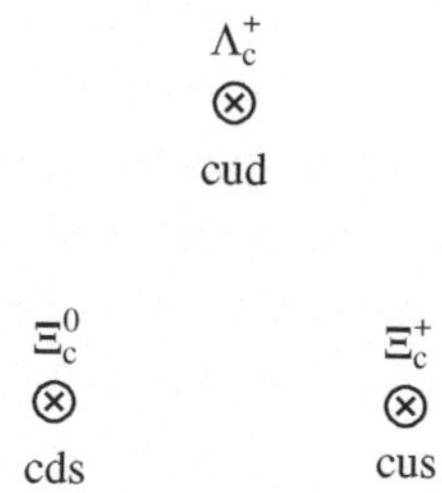

Fig. 2.2. Flavor $SU(3)$ weight diagram for the $\bar{\mathbf{3}}$ spin-$1/2$ $c\,[qq]$ baryons. The corresponding $b\,[qq]$ baryons are the Λ_b^0, Ξ_b^- and Ξ_b^0. The vertical direction is hypercharge, and the horizontal direction is I_3, the third component of isospin.

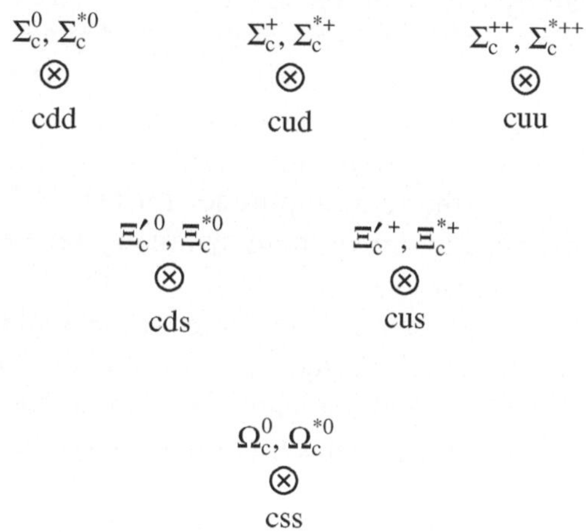

Fig. 2.3. Flavor $SU(3)$ weight diagram for the $\mathbf{6}$ spin-$1/2$ and spin-$3/2$ $c\,(qq)$ baryons. The corresponding $b\,(qq)$ baryons are the spin-$1/2$ $\Sigma_b^{-,0,+}$, $\Xi_b'^{-,0}$, and Ω_b^-, and their spin-$3/2$ partners. The vertical direction is hypercharge, and the horizontal direction is I_3, the third component of isospin.

constituent quark model. In this model the ground-state baryons have no orbital angular momentum and the spatial wave function for the two light constituent quarks is symmetric under their interchange. The wave function is also completely antisymmetric in color. Fermi statistics then demands that for $s_\ell = 0$, where the spin wave function is antisymmetric, the $SU(3)_V$ flavor wave function is also antisymmetric, and hence transforms as $(\mathbf{3} \times \mathbf{3})_{\text{antisymmetric}} = \bar{\mathbf{3}}$. For $s_\ell = 1$, the $SU(3)_V$ flavor wave function is symmetric and hence transforms as $(\mathbf{3} \times \mathbf{3})_{\text{symmetric}} = \mathbf{6}$. The $s_\ell = 0$ ground-state baryons have positive parity and total spin of $1/2$, and the spinor fields that destroy these states are denoted by $\Lambda_v^{(Q)}$. The $s_\ell = 1$ ground-state baryons have positive parity and come in a doublet of states with total spins of $1/2$, and $3/2$. We denote the fields that destroy these states by $\Sigma_v^{(Q)}$ and $\Sigma_{v\mu}^{*(Q)}$, respectively. The spectrum of excited baryons is more complicated than in the meson sector. In the nonrelativistic constituent quark model, the $L = 1$ baryons come in two types; states with the unit of orbital angular momentum between the two light quarks, and states with the unit of orbital angular momentum between the light quark pair and the heavy quark. The latter are expected to be lower in mass. The lowest-mass hadrons containing c and b quarks are summarized in Tables 2.1 and 2.2, respectively.

2.3 Strong decays of excited heavy hadrons

In many cases the two members of a doublet with spin of the light degrees of freedom s_ℓ can decay by means of a single pion emission to the two members of another lower-mass doublet with spin of the light degrees of freedom s'_ℓ. The orbital angular momentum of the emitted pion (L, L_z) is restricted by parity, angular momentum conservation, and heavy quark spin symmetry. For a given pion partial wave there are four transition amplitudes that are related by heavy quark spin symmetry, e.g., the four amplitudes for $(D_1, D_2^*) \to (D, D^*) + \pi$. It is an instructive exercise to derive these symmetry relations. The derivation only makes use of the standard formula for the addition of angular momenta in quantum mechanics. The first step is to decompose the total angular momentum of the initial and final heavy hadron states j and j' into the spin of the initial and final heavy quark s_Q and s'_Q, and the spin of the initial and final light degrees of freedom s_ℓ and s'_ℓ. Using the Clebsch-Gordan decomposition of $|j, j_z\rangle$ into $|\frac{1}{2}, s_{Q_z}\rangle$ and $|s_\ell, s_{\ell_z}\rangle$,

$$|j, j_z\rangle = \sum_{s_{Qz}, s_{\ell z}} \left\langle \tfrac{1}{2}, s_{Q_z}; s_\ell, s_{\ell_z} \middle| \, j, j_z \right\rangle \left| \tfrac{1}{2}, s_{Q_z} \right\rangle \left| s_\ell, s_{\ell_z} \right\rangle, \tag{2.2}$$

and the corresponding decomposition of $|j', j'_z\rangle$ into $|\frac{1}{2}, s'_{Q_z}\rangle$ and $|s'_\ell, s'_{\ell_z}\rangle$, the transition amplitude can be written in the form

$$\begin{aligned}
&\mathcal{M}\left[H^{(Q)}(j, j_z) \to H^{(Q)}(j', j'_z) + \pi(L, L_z)\right] \\
&\quad = \langle \pi\,(L, L_z);\; j', j'_z | H_{\text{eff}} | j, j_z \rangle \\
&\quad = \sum \left\langle \pi\,(L, L_z);\; \tfrac{1}{2}, s'_{Qz};\; s'_\ell, s'_{\ell z} \middle| \, H_{\text{eff}} \, \middle| \tfrac{1}{2}, s_{Qz};\; s_\ell, s_{\ell z} \right\rangle \\
&\qquad \times \left\langle \tfrac{1}{2}, s'_{Qz}; s'_\ell, s'_{\ell z} \middle| \, j', j'_z \right\rangle \left\langle \tfrac{1}{2}, s_{Qz}; s_\ell, s_{\ell z} \middle| \, j, j_z \right\rangle.
\end{aligned} \tag{2.3}$$

Eq. (2.3) is schematic and only keeps track of the group theory factors. The effective strong interaction Hamiltonian, H_{eff}, conserves the spin of the heavy quark and of the light degrees of freedom separately. The Wigner-Eckart theorem then implies that the hadronic matrix element must have the form

$$\begin{aligned}
&\left\langle \pi\,(L, L_z);\; \tfrac{1}{2}, s'_{Qz};\; s'_\ell, s'_{\ell z} \middle| \, H_{\text{eff}} \, \middle| \tfrac{1}{2}, s_{Qz};\; s_\ell, s_{\ell z} \right\rangle \\
&\qquad = \delta_{s_{Qz}, s'_{Qz}} \langle L, L_z; s'_\ell, s'_{\ell z} | s_\ell, s_{\ell z} \rangle \langle L, s'_\ell \parallel H_{\text{eff}} \parallel s_\ell \rangle,
\end{aligned} \tag{2.4}$$

Table 2.1. *The lowest-mass hadrons containing a c quark*[a]

Hadron	Mass (MeV)	Quark Content	J^P	s_ℓ
D^+	1869.3 ± 0.5	$c\bar{d}$	0^-	1/2
D^{*+}	2010.0 ± 0.5		1^-	
D^0	1864.6 ± 0.5	$c\bar{u}$	0^-	1/2
D^{*0}	2006.7 ± 0.5		1^-	
D_s^+	1968.5 ± 0.6	$c\bar{s}$	0^-	1/2
D_s^{*+}	2112.4 ± 0.7		1^-	
D_0^*		$c\bar{q}$	0^+	1/2
D_1^*	2461 ± 50		1^+	
D_1	2422.2 ± 1.8	$c\bar{q}$	1^+	3/2
D_2^*	2458.9 ± 2.0		2^+	
Λ_c^+	2284.9 ± 0.6	$c[ud]$	$1/2^+$	0
Ξ_c^+	2465.6 ± 1.4	$c[us]$	$1/2^+$	0
Ξ_c^0	2470.3 ± 1.8	$c[ds]$	$1/2^+$	0
Σ_c^{++}	2452.8 ± 0.6	$c(uu)$	$1/2^+$	1
Σ_c^{*++}	2519.4 ± 1.5		$3/2^+$	
Σ_c^+	2453.6 ± 0.9	$c(ud)$	$1/2^+$	1
Σ_c^{*+}			$3/2^+$	
Σ_c^0	2452.2 ± 0.6	$c(dd)$	$1/2^+$	1
Σ_c^{*0}	2517.5 ± 1.4		$3/2^+$	
$\Xi_c'^+$	2573.4 ± 3.3	$c(us)$	$1/2^+$	1
Ξ_c^{*+}	2644.6 ± 2.1		$3/2^+$	
$\Xi_c'^0$	2577.3 ± 3.4	$c(ds)$	$1/2^+$	1
Ξ_c^{*0}	2643.8 ± 1.8		$3/2^+$	
Ω_c^0	2704 ± 4	$c(ss)$	$1/2^+$	1
Ω_c^{*0}			$3/2^+$	

[a] Heavy quark spin symmetry multiplets are listed together. For the excited mesons, the masses quoted correspond to $q = u, d$. Excited charm masses with quark content $c\bar{s}$ and excited charm baryons have also been observed.

Table 2.2. *The lowest-mass hadrons containing a b quark*[a]

Hadron	Mass (MeV)	Quark Content	J^P	s_ℓ
$\bar{B}^0$	5279.2 ± 1.8	$b\bar{d}$	0^-	1/2
$\bar{B}^{*0}$	5324.9 ± 1.8		1^-	
$\bar{B}^-$	5278.9 ± 1.8	$b\bar{u}$	0^-	1/2
$\bar{B}^{*-}$	5324.9 ± 1.8		1^-	
$\bar{B}_s^0$	5369.3 ± 2.0	$b\bar{s}$	0^-	1/2
$\bar{B}_s^{*0}$			1^-	
$\bar{B}_0^*$		$b\bar{q}$	0^+	1/2
$\bar{B}_1^*$			1^+	
$\bar{B}_1$		$b\bar{q}$	1^+	3/2
$\bar{B}_2^*$			2^+	
Λ_b^0	5624 ± 9	$b[ud]$	$1/2^+$	0
Ξ_b^0		$b[us]$	$1/2^+$	0
Ξ_b^-		$b[ds]$	$1/2^+$	0
Σ_b^+		$b(uu)$	$1/2^+$	1
Σ_b^{*+}			$3/2^+$	
Σ_b^0		$b(ud)$	$1/2^+$	1
Σ_b^{*0}			$3/2^+$	
Σ_b^-		$b(dd)$	$1/2^+$	1
Σ_b^{*-}			$3/2^+$	
$\Xi_b^{\prime 0}$		$b(us)$	$1/2^+$	1
Ξ_b^{*0}			$3/2^+$	
$\Xi_b^{\prime -}$		$b(ds)$	$1/2^+$	1
Ξ_b^{*-}			$3/2^+$	
Ω_b^-		$b(ss)$	$1/2^+$	1
Ω_b^{*-}			$3/2^+$	

[a] Heavy quark spin symmetry multiplets are listed together.

where the final term is the reduced matrix element. Substituting into Eq. (2.3) yields

$$\begin{aligned}
\mathcal{M} &= \sum \left\langle \tfrac{1}{2}, s_{Qz}; s_\ell, s_{\ell z} \right| j, j_z \rangle \langle L, s'_\ell \parallel H_{\text{eff}} \parallel s_\ell \rangle \\
&\quad \times \left\langle \tfrac{1}{2}, s_{Qz}; s'_\ell, s'_{\ell z} \right| j', j'_z \rangle \left\langle L, L_z; s'_\ell, s'_{\ell z} \right| s_\ell, s_{\ell z} \rangle \\
&= (-1)^{L+s'_\ell+\frac{1}{2}+j} \sqrt{(2s_\ell+1)(2j'+1)} \begin{Bmatrix} L & s'_\ell & s_\ell \\ \frac{1}{2} & j & j' \end{Bmatrix} \\
&\quad \times \langle L, (j_z - j'_z); j', j'_z | j, j_z \rangle \langle L, s'_\ell \parallel H_{\text{eff}} \parallel s_\ell \rangle,
\end{aligned} \tag{2.5}$$

where we have rewritten the product of Clebsch-Gordan coefficients in terms of $6j$ symbols. The total decay rate for $j \to j'$ is given by

$$\begin{aligned}
\Gamma(j \to j'\pi) &\propto (2s_\ell+1) \frac{2j'+1}{2j+1} \sum_{j_z, j'_z} \left| \begin{Bmatrix} L & s'_\ell & s_\ell \\ \frac{1}{2} & j & j' \end{Bmatrix} \right|^2 |\langle L, (j_z - j'_z); j', j'_z | j, j_z \rangle|^2 \\
&= (2s_\ell+1)(2j'+1) \left| \begin{Bmatrix} L & s'_\ell & s_\ell \\ \frac{1}{2} & j & j' \end{Bmatrix} \right|^2,
\end{aligned} \tag{2.6}$$

where we have dropped terms, such as the reduced matrix element, which are the same for different values of j and j'. Equation (2.6) provides relations between the decay rates of the excited $s_\ell = 3/2$ D_1 and D_2^* mesons to the ground state $s_\ell = 1/2$ D or D^* mesons and a pion. These two multiplets have opposite parity and the pion has negative parity, so the pion must be in an even partial wave with $L = 0$ or 2 by parity and angular momentum conservation. The decays $D_2^* \to D\pi$ and $D_2^* \to D^*\pi$ must occur through the $L = 2$ partial wave, while $D_1 \to D^*\pi$ can occur by either the $L = 0$ or $L = 2$ partial wave. The $L = 0$ partial wave amplitude for $D_1 \to D^*\pi$ vanishes by heavy quark symmetry since

$$\begin{Bmatrix} 0 & 1/2 & 3/2 \\ 1/2 & 1 & 1 \end{Bmatrix} = 0, \tag{2.7}$$

so that all the decays are $L = 2$. Equation (2.6) implies that the $L = 2$ decay rates are in the ratio

$$\begin{matrix} \Gamma(D_1 \to D\pi) : \Gamma(D_1 \to D^*\pi) : \Gamma(D_2^* \to D\pi) : \Gamma(D_2^* \to D^*\pi) \\ 0 \quad : \quad 1 \quad : \quad \frac{2}{5} \quad : \quad \frac{3}{5} \end{matrix}, \tag{2.8}$$

where $\Gamma(D_1 \to D\pi)$ is forbidden by angular momentum and parity conservation. Equation (2.8) holds in the heavy quark symmetry limit, $m_c \to \infty$. There is a very important source of heavy quark spin symmetry violation that is kinematic in origin. For small $\mathbf{p}_\pi$, the decay rates are proportional to $|\mathbf{p}_\pi|^{2L+1}$, which for $L = 2$ is $|\mathbf{p}_\pi|^5$. In the $m_c \to \infty$ limit the D_1 and D_2^* are degenerate and the D and D^* are also degenerate. Consequently this factor does not affect the ratios in Eq. (2.8). However, for the physical value of m_c, the $D^* - D$

mass splitting is ~140 MeV, which cannot be neglected in comparison with the 450 MeV $D_2^* - D^*$ splitting. Including the factor of $|\mathbf{p}_\pi|^5$, the relative decay rates become

$$\begin{array}{ccccccc} \Gamma(D_1 \to D\pi) & : & \Gamma(D_1 \to D^*\pi) & : & \Gamma(D_2^* \to D\pi) & : & \Gamma(D_2^* \to D^*\pi) \\ 0 & : & 1 & : & 2.3 & : & 0.92 \end{array} . \tag{2.9}$$

As a consequence of Eq. (2.9) we arrive at the prediction BR$(D_2^* \to D\pi)$/BR $(D_2^* \to D^*\pi) \simeq 2.5$, which is in good agreement with the experimental value 2.3 ± 0.6. The prediction for this ratio of branching ratios would have been 2/3 without including the phase space correction factor.

Phenomenologically, the suppression associated with emission of a low-momentum pion in a partial wave L is $\sim(|\mathbf{p}_\pi|/\Lambda_{\rm CSB})^{2L+1}$. The fact that the scale $\Lambda_{\rm CSB} \sim 1$ GeV enables us to understand why the doublet of D_0^* and D_1^* excited $s_\ell = 1/2$ mesons is difficult to observe. For these mesons, heavy quark spin symmetry predicts that their decays to the ground-state doublet by single pion emission occur in the $L = 0$ partial wave. The masses of the $(D_0^*,\ D_1^*)$ are expected to be near the masses of the $(D_1,\ D_2^*)$, and so their widths are larger than those of the D_1 and D_2^* by roughly $(\Lambda_{\rm CSB}/|\mathbf{p}_\pi|)^4 \sim 20$–40. The D_1 and D_2^* widths are $\Gamma(D_1) = 18.9 \pm 4$ MeV and $\Gamma(D_2^*) = 23 \pm 5$ MeV. Hence the $D_{0,1}^*$ should be broad, with widths greater than 200 MeV, which makes them difficult to observe. The measured width of the D_1^* is 290 ± 100 MeV.

The excited positive parity $s_\ell = 3/2$ mesons D_{s1} and D_{s2}^*, which contain a strange antiquark, have also been observed. The D_{s1} is narrow, $\Gamma(D_{s1}) <$ 2.3 MeV, and its decays to D^*K are dominated by the S-wave amplitude. This occurs because the kaon mass is much larger than the pion mass, and so for this decay $|\mathbf{p}_K| \simeq 150$ MeV while in $D_1 \to D^*\pi$ decay $|\mathbf{p}_\pi| \simeq 360$ MeV. Consequently, there is a large kinematic suppression of the D-wave amplitude in $D_{s1} \to D^*K$ decay. The $s_\ell = 1/2$ and $s_\ell = 3/2$ charmed mesons are in a $\bar{\mathbf{3}}$ of $SU(3)_V$, whereas the π, K, and η are in an **8**. Since there is only one way to combine a **3**, $\bar{\mathbf{3}}$, and **8** into a singlet, $SU(3)_V$ relates the S-wave part of the D_1 decay width to the D_{s1} decay width. Neglecting η final states, which are phase space suppressed, $SU(3)_V$ light quark symmetry leads to the expectation that $\Gamma_{S-{\rm wave}}(D_1) \approx (3/4)\Gamma(D_{s1}) \times |\mathbf{p}_\pi|/|\mathbf{p}_K| < 4.1$ MeV.

2.4 Fragmentation to heavy hadrons

A heavy quark produced in a high-energy process will materialize as a hadron containing that heavy quark. Once the "off-shellness" of the fragmenting heavy quark is small compared with its mass, the fragmentation process is constrained by heavy quark symmetry. Heavy quark symmetry implies that the probability, $P_{h_Q \to h_s}^{(H)}$, for a heavy quark Q with spin along the fragmentation axis (i.e., helicity) h_Q to fragment to a hadron H with spin s, spin of the light degrees of freedom s_ℓ, and helicity h_s is

$$P^{(H)}_{h_Q \to h_s} = \sum_{h_\ell} P_{Q \to s_\ell}\, p_{h_\ell} |\langle s_Q, h_Q; s_\ell, h_\ell | s, h_s\rangle|^2, \tag{2.10}$$

where $h_\ell = h_s - h_Q$. In Eq. (2.10) $P_{Q \to s_\ell}$ is the probability for the heavy quark to fragment to a hadron with spin of the light degrees of freedom s_ℓ. This probability is independent of the spin and flavor of the heavy quark but will depend on other quantum numbers needed to specify the hadron H. $P_{Q \to s_\ell}$ has the same value for the two hadrons in the doublet related by heavy quark spin symmetry. p_{h_ℓ} is the conditional probability that the light degrees of freedom have helicity h_ℓ, given that Q fragments to s_ℓ. The probabilistic interpretation of the fragmentation process means that $0 \le p_{h_\ell} \le 1$ and

$$\sum_{h_\ell} p_{h_\ell} = 1. \tag{2.11}$$

Like $P_{Q \to s_\ell}$, p_{h_ℓ} is independent of the spin and flavor of the heavy quark, but can depend on the hadron multiplet. The third factor in Eq. (2.10) is the Clebsch-Gordan probability that the hadron H with helicity h_s contains light degrees of freedom with helicity h_ℓ and a heavy quark with helicity h_Q. Parity invariance of the strong interactions implies that

$$p_{h_\ell} = p_{-h_\ell}, \tag{2.12}$$

since reflection in a plane containing the momentum of the fragmenting quark reverses the helicities but leaves the momentum unchanged. Equations (2.11) and (2.12) imply that the number of independent probabilities p_{h_ℓ} is $s_\ell - 1/2$ for mesons and s_ℓ for baryons. At the hadron level, parity invariance of the strong interactions gives the relation $P^{(H)}_{h_Q \to h_s} = P^{(H)}_{-h_Q \to -h_s}$.

Heavy quark spin symmetry has reduced the number of independent fragmentation probabilities. For the ground-state D and D^* mesons, $s_\ell = 1/2$, so $p_{1/2} = p_{-1/2}$, which must both equal 1/2, since $p_{1/2} + p_{-1/2} = 1$. This gives the relative fragmentation probabilities for a right-handed charm quark,

$$\begin{array}{ccccccc} P^{(D)}_{1/2 \to 0} & : & P^{(D^*)}_{1/2 \to 1} & : & P^{(D^*)}_{1/2 \to 0} & : & P^{(D^*)}_{1/2 \to -1} \\ 1/4 & : & 1/2 & : & 1/4 & : & 0 \end{array}. \tag{2.13}$$

Parity invariance of the strong interactions relates the fragmentation probabilities for a left-handed charm quark to those in Eq. (2.13). Heavy quark spin symmetry implies that a charm quark fragments to a D one-third as often as it fragments to a D^*. This prediction disagrees with the experimental data, which give a larger fragmentation probability for the D, and the discrepancy is due to the $D^* - D$ mass difference. We have already seen that the mass difference has an important impact on decays of excited charm mesons to the D and D^* and it is not surprising that the mass difference should influence the fragmentation probabilities as well. The $B^* - B$ mass difference is 50 MeV, which is approximately a factor of 3 smaller than the $D^* - D$ mass difference, so one expects the predictions of exact heavy quark symmetry to work better in this case. Recent experimental data from LEP show that the $B^* : B$ ratio is consistent with the predicted value of 3 : 1.

Charm quark fragmentation to the negative parity $s_\ell = 3/2$ multiplet of excited charmed mesons is characterized by the Falk-Peskin parameter $w_{3/2}$, defined as the conditional probability to fragment to helicities $\pm 3/2$,

$$p_{3/2} = p_{-3/2} = \tfrac{1}{2} w_{3/2}, \qquad p_{1/2} = p_{-1/2} = \tfrac{1}{2}(1 - w_{3/2}). \tag{2.14}$$

The value of $p_{\pm 1/2}$ is determined in terms of $w_{3/2}$ since the total fragmentation probability must be unity. The relative fragmentation probabilities are given by Eq. (2.10):

$$\begin{array}{ccccccc} P^{(D_1)}_{1/2\to 1} & : & P^{(D_1)}_{1/2\to 0} & : & P^{(D_1)}_{1/2\to -1} & : & P^{(D_2^*)}_{1/2\to 2} \; : \\ \frac{1}{8}(1-w_{3/2}) & : & \frac{1}{4}(1-w_{3/2}) & : & \frac{3}{8}w_{3/2} & : & \frac{1}{2}w_{3/2} \; : \\ P^{(D_2^*)}_{1/2\to 1} & : & P^{(D_2^*)}_{1/2\to 0} & : & P^{(D_2^*)}_{1/2\to -1} & : & P^{(D_2^*)}_{1/2\to -2} \\ \frac{3}{8}(1-w_{3/2}) & : & \frac{1}{4}(1-w_{3/2}) & : & \frac{1}{8}w_{3/2} & : & 0 \end{array} . \tag{2.15}$$

Equation (2.15) predicts that the ratio of D_1 to D_2^* production by charm quark fragmentation is 3/5, independent of $w_{3/2}$. Assuming that the decays of the negative parity $s_\ell = 3/2$ charmed mesons are dominated by $D^{(*)}\pi$ final states, the experimental value of this ratio is close to unity. Experimentally the probability of a heavy quark to fragment to the maximal helicities $\pm 3/2$ is small, i.e., $w_{3/2} < 0.24$.

The validity of Eq. (2.10) depends on a crucial assumption. Spin symmetry violation must be negligible in the masses and decays of excited multiplets that can be produced in the fragmentation process and then decay to the final fragmentation product. The spin symmetry violating $D_1 - D_2^*$ mass difference is comparable with the widths of these states, and the spin symmetry violating $D^* - D$ mass difference plays an important role in their decay rates to D and D^*'s. Consequently we do not expect Eq. (2.13) to hold for those D and D^*'s that arise from decays of a D_1 or D_2^*.

2.5 Covariant representation of fields

We have seen that heavy quark symmetry usually implies a degenerate multiplet of states, such as the B and B^*. It is convenient to have a formalism in which the entire multiplet of degenerate states is treated as a single object that transforms linearly under the heavy quark symmetries.

The ground-state $Q\bar{q}$ mesons can be represented by a field $H_v^{(Q)}$ that annihilates the mesons, and transforms as a bilinear under Lorentz transformations,

$$H_{v'}^{(Q)\prime}(x') = D\,(\Lambda)\, H_v^{(Q)}(x)\, D\,(\Lambda)^{-1}\,, \tag{2.16}$$

where

$$v' = \Lambda v, \qquad x' = \Lambda x, \tag{2.17}$$

and $D\,(\Lambda)$ is the Lorentz transformation matrix for spinors, so that

$$H_v^{(Q)}\,(x) \to H_v^{(Q)\prime}(x) = D\,(\Lambda)\, H_{\Lambda^{-1}v}^{(Q)}(\Lambda^{-1}x) D\,(\Lambda)^{-1}\,. \tag{2.18}$$

The field $H_v^{(Q)}(x)$ is a linear combination of the pseudoscalar field $P_v^{(Q)}(x)$ and the vector field $P_{v\mu}^{*(Q)}\,(x)$ that annihilate the $s_\ell = 1/2$ meson multiplet. Vector particles have a polarization vector ϵ_μ, with $\epsilon \cdot \epsilon = -1$, and $v \cdot \epsilon = 0$. The amplitude for $P_{v\mu}^{*(Q)}$ to annihilate a vector particle is ϵ_μ. A simple way to combine the two fields into a single field with the desired transformation properties is to define*

$$H_v^{(Q)} = \frac{1 + \not{v}}{2}\left[\not{P}_v^{*(Q)} + i\, P_v^{(Q)} \gamma_5\right]. \tag{2.19}$$

Equation (2.19) is consistent with $P_v^{(Q)}$ transforming as a pseudoscalar, and $P_{v\mu}^{*(Q)}$ as a vector, since γ_5 and γ^μ convert pseudoscalars and vectors into bispinors. The $\left(1 + \not{v}\right)/2$ projector retains only the particle components of the heavy quark Q. The relative sign and phase between the P and P^* terms in Eq. (2.19) is arbitrary, and this depends on the choice of phase between the pseudoscalar and vector meson states. The pseudoscalar is multiplied by γ_5 rather than unity, to be consistent with the parity transformation law

$$H_v^{(Q)}\,(x) \to \gamma^0 H_{v_P}^{(Q)}\,(x_P)\, \gamma^0, \tag{2.20}$$

where

$$x_P = (x^0, -\mathbf{x}), \qquad v_P = (v^0, -\mathbf{v}). \tag{2.21}$$

The field $H_v^{(Q)}$ satisfies the constraints

$$\not{v} H_v^{(Q)} = H_v^{(Q)}, \qquad H_v^{(Q)} \not{v} = -H_v^{(Q)}. \tag{2.22}$$

The first of these follows directly from $\not{v}\left(1 + \not{v}\right) = \left(1 + \not{v}\right)$. The second relation follows by anticommuting $\not{v}$ through $H_v^{(Q)}$, and using $v \cdot P_v^{*(Q)} = 0$, since the polarization of physical spin-one particles satisfies $v \cdot \epsilon = 0$.

* For clarity, the superscript (Q) and/or the subscript v will sometimes be omitted.

It is convenient to introduce the conjugate field

$$\bar{H}_v^{(Q)} = \gamma^0 H_v^{(Q)\dagger}\gamma^0 = \left[P_{v\mu}^{*(Q)\dagger}\gamma^\mu + iP_v^{(Q)\dagger}\gamma_5\right]\frac{1+\not{v}}{2}, \tag{2.23}$$

which also transforms as a bispinor,

$$\bar{H}_v^{(Q)}(x) \to D(\Lambda)\,\bar{H}_{\Lambda^{-1}v}^{(Q)}(\Lambda^{-1}x)D(\Lambda)^{-1}\,, \tag{2.24}$$

since

$$\gamma^0 D(\Lambda)^\dagger\,\gamma^0 = D(\Lambda)^{-1}\,. \tag{2.25}$$

In the rest frame

$$v = v_r = (1, \mathbf{0}) \tag{2.26}$$

the field $H_{v_r}^{(Q)}$ is

$$H_{v_r}^{(Q)} = \begin{pmatrix} 0 & iP_{v_r}^{(Q)} - \boldsymbol{\sigma}\cdot\mathbf{P}_{v_r}^{*(Q)} \\ 0 & 0 \end{pmatrix}, \tag{2.27}$$

using the Bjorken and Drell convention for γ matrices,

$$\gamma^0 = \begin{pmatrix} 1 & 0 \\ 0 & -1 \end{pmatrix}, \qquad \boldsymbol{\gamma} = \begin{pmatrix} 0 & \boldsymbol{\sigma} \\ -\boldsymbol{\sigma} & 0 \end{pmatrix}, \qquad \gamma_5 = \begin{pmatrix} 0 & 1 \\ 1 & 0 \end{pmatrix}. \tag{2.28}$$

The indices α and β of the field $[H_{v_r}^{(Q)}]_{\alpha\beta}$ label the spinor indices of the heavy quark Q and the light degrees of freedom, respectively. The field $H_{v_r}^{(Q)}$ transforms as a $(1/2, 1/2)$ representation under $S_Q \otimes S_\ell$. The spin operators $\mathbf{S_Q}$ and $\mathbf{S}_\ell$ for the heavy quark and light degrees of freedom acting on the $H_{v_r}^{(Q)}$ field are

$$\begin{aligned} \left[\mathbf{S_Q}, H_{v_r}^{(Q)}\right] &= \tfrac{1}{2}\boldsymbol{\sigma}_{\mathbf{4}\times\mathbf{4}} H_{v_r}^{(Q)}, \\ \left[\mathbf{S}_\ell, H_{v_r}^{(Q)}\right] &= -\tfrac{1}{2} H_{v_r}^{(Q)}\boldsymbol{\sigma}_{\mathbf{4}\times\mathbf{4}}, \end{aligned} \tag{2.29}$$

where $\sigma_{4\times4}^i = i\epsilon_{ijk}[\gamma^j, \gamma^k]/4$ are the usual Dirac rotation matrices in the spinor representation. Under infinitesimal rotations, one finds (neglecting derivative terms that arise from rotating the spatial dependence of the fields) that

$$\delta H_{v_r}^{(Q)} = i\left[\boldsymbol{\theta}\cdot(\mathbf{S}_Q + \mathbf{S}_\ell), H_{v_r}^{(Q)}\right] = \frac{i}{2}\left[\boldsymbol{\theta}\cdot\boldsymbol{\sigma}_{\mathbf{4}\times\mathbf{4}}, H_{v_r}^{(Q)}\right], \tag{2.30}$$

so that

$$\delta P_{v_r}^{(Q)} = 0, \qquad \delta\mathbf{P}_{v_r}^{*(Q)} = \boldsymbol{\theta}\times\mathbf{P}_{v_r}^{*(Q)}, \tag{2.31}$$

which are the transformation rules for a spin-zero and spin-one particle, respectively. The fields $P_v^{(Q)}$ and $P_{v\mu}^{*(Q)}$ mix under S_Q or S_ℓ transformations. Under heavy quark spin transformations,

$$\delta H_{v_r}^{(Q)} = i\left[\boldsymbol{\theta}\cdot\mathbf{S_Q},\, H_{v_r}^{(Q)}\right] = \frac{i}{2}\boldsymbol{\theta}\cdot\boldsymbol{\sigma}_{4\times 4}H_{v_r}^{(Q)}, \tag{2.32}$$

so that

$$\delta P_{v_r}^{(Q)} = -\tfrac{1}{2}\boldsymbol{\theta}\cdot\mathbf{P}_{v_r}^{*(Q)}, \qquad \delta\mathbf{P}_{v_r}^{*(Q)} = \tfrac{1}{2}\boldsymbol{\theta}\times\mathbf{P}_{v_r}^{*(Q)} - \tfrac{1}{2}\boldsymbol{\theta}P_{v_r}^{(Q)}. \tag{2.33}$$

Under finite heavy quark spin transformations,

$$H_v^{(Q)} \to D(R)_Q H_v^{(Q)}, \tag{2.34}$$

where $D(R)_Q$ is the rotation matrix in the spinor representation for the rotation R. Like the Lorentz transformations, it satisfies $\gamma^0 D(R)_Q^\dagger\gamma^0 = D(R)_Q^{-1}$.

It is straightforward to write couplings that are invariant under the heavy quark symmetry using the field $H_v^{(Q)}$ and its transformation rules. We have concentrated on the heavy quark spin symmetry, because that is the new ingredient in the formalism. One can also implement the heavy quark flavor symmetry by using fields $H_v^{(Q_i)}$ for each heavy quark flavor Q_i, and also imposing heavy flavor symmetry

$$H_v^{(Q_i)} \to U_{ij}H_v^{(Q_j)}, \tag{2.35}$$

where U_{ij} is an arbitrary unitary matrix in flavor space.

We have seen how to use a covariant formalism for the pseudoscalar and vector meson multiplet. It is straightforward to derive a similar formalism for baryon states. For example the Λ_Q baryon has light degrees of freedom with spin zero, so the spin of the baryon is the spin of the heavy quark. It is described by a spinor field $\Lambda_v^{(Q)}(x)$ that satisfies the constraint

$$\not{v}\Lambda_v^{(Q)} = \Lambda_v^{(Q)}, \tag{2.36}$$

transforms under the Lorentz group as†

$$\Lambda_v^{(Q)}(x) \to D(\Lambda)\,\Lambda_{\Lambda^{-1}v}^{(Q)}(\Lambda^{-1}x), \tag{2.37}$$

and transforms under heavy quark spin transformations as

$$\Lambda_v^{(Q)} \to D(R)_Q\Lambda_v^{(Q)}. \tag{2.38}$$

The analog of the polarization vector for spin-1/2 Λ_Q states with velocity v and spin s is the spinor $u(v, s)$. These spinors will be normalized so that

$$\bar{u}(v, s)\gamma^\mu u(v, s) = 2v^\mu. \tag{2.39}$$

† We hope the reader is not confused by the use of Λ for both the Lorentz transformation and the heavy baryon field.

Then

$$\bar{u}(v,s)\gamma^{\mu}\gamma_5 u(v,s) = 2s^{\mu}, \tag{2.40}$$

where s^{μ} is the spin vector, satisfying $v \cdot s = 0$ and $s^2 = -1$. The field $\Lambda_v^{(Q)}$ annihilates heavy baryon states with amplitude $u(v,s)$.

2.6 The effective Lagrangian

The QCD Lagrangian does not have manifest heavy quark spin-flavor symmetry as $m_Q \to \infty$. It is convenient to use an effective field theory for QCD in which heavy quark symmetry is manifest in the $m_Q \to \infty$ limit. This effective field theory is known as heavy quark effective theory (HQET), and it describes the dynamics of hadrons containing a single heavy quark. It is a valid description of the physics at momenta much smaller than the mass of the heavy quark m_Q. The effective field theory is constructed so that only inverse powers of m_Q appear in the effective Lagrangian, in contrast to the QCD Lagrangian in Eq. (1.82), which has positive powers of m_Q.

Consider a single heavy quark with velocity v interacting with external fields, where the velocity of an on-shell quark is defined by $p = m_Q v$. The momentum of an off-shell quark can be written as $p = m_Q v + k$, where the residual momentum k determines the amount by which the quark is off shell because of its interactions. For heavy quarks in a hadron, k is of the order of $\Lambda_{\rm QCD}$. The usual Dirac quark propagator simplifies to

$$i\frac{\not{p} + m_Q}{p^2 - m_Q^2 + i\varepsilon} = i\frac{m_Q \not{v} + m_Q + \not{k}}{2m_Q v \cdot k + k^2 + i\varepsilon} \to i\frac{1 + \not{v}}{2v \cdot k + i\varepsilon} \tag{2.41}$$

in the heavy quark limit. The propagator contains a velocity-dependent projection operator

$$\frac{1 + \not{v}}{2}. \tag{2.42}$$

In the rest frame of the heavy quark this projection operator becomes $(1 + \gamma^0)/2$, which projects onto the particle components of the four-component Dirac spinor.

It is convenient to formulate the effective Lagrangian directly in terms of velocity-dependent fields $Q_v(x)$, which are related to the original quark fields $Q(x)$ at tree level. One can write the original quark field $Q(x)$ as

$$Q(x) = e^{-im_Q v \cdot x}\left[Q_v(x) + \mathfrak{Q}_v(x)\right], \tag{2.43}$$

where

$$Q_v(x) = e^{im_Q v \cdot x}\frac{1 + \not{v}}{2}Q(x), \qquad \mathfrak{Q}_v(x) = e^{im_Q v \cdot x}\frac{1 - \not{v}}{2}Q(x). \tag{2.44}$$

The exponential prefactor subtracts $m_Q v^\mu$ from the heavy quark momentum. The Q_v field produces effects at leading order, whereas the effects of $\mathfrak{Q}_v$ are suppressed by powers of $1/m_Q$. These $1/m_Q$ corrections are discussed in Chapter 4. Neglecting $\mathfrak{Q}_v$ and substituting Eq. (2.43) into the part of the QCD Lagrangian density involving the heavy quark field, $\bar{Q}(i\not{D} - m_Q)Q$ gives $\bar{Q}_v i\not{D} Q_v$. Inserting $(1+\not{v})/2$ on either side of $\not{D}$ yields

$$\mathcal{L} = \bar{Q}_v \, (iv \cdot D) \, Q_v, \tag{2.45}$$

which is an m_Q-independent expression. The Q_v propagator that follows from Eq. (2.45) is

$$\left(\frac{1+\not{v}}{2}\right) \frac{i}{(v \cdot k + i\varepsilon)}, \tag{2.46}$$

which is the same as was derived previously by taking the $m_Q \to \infty$ limit of the Feynman rules. The projector in Eq. (2.46) arises because Q_v satisfies

$$\left(\frac{1+\not{v}}{2}\right) Q_v = Q_v. \tag{2.47}$$

Beyond tree level, there is no simple connection between the fields Q_v of the effective Lagrangian and Q of the QCD theory. The effective theory is constructed by making sure that on-shell Green's functions in the effective theory are equal to those in QCD to a given order in $1/m_Q$ and $\alpha_s(m_Q)$. At tree level, we have seen that the quark propagator in the effective theory matches that in the full theory up to terms of the order of $1/m_Q$. It remains to show that the gluon interaction vertex is the same in the two theories. Consider a generic gluon interaction, as shown in Fig. 2.4. The interaction vertex in the full theory is $-igT^A\gamma^\mu$, whereas in the effective theory, the vertex is $-igT^A v^\mu$ from the $v \cdot D$ term in Eq. (2.45). The vertex in the full theory is sandwiched between quark propagators. Each heavy quark propagator is proportional to $(1+\not{v})/2$, so the factor of γ^μ in the vertex can be replaced by

$$\gamma^\mu \to \frac{1+\not{v}}{2}\gamma^\mu\frac{1+\not{v}}{2} = v^\mu\frac{1+\not{v}}{2} \to v^\mu, \tag{2.48}$$

which gives the same vertex as in the effective theory. Thus the effective Lagrangian in Eq. (2.45) reproduces all the Green's functions in the full theory to leading order in $1/m_Q$ and $\alpha_s(m_Q)$. If there is more than one heavy quark

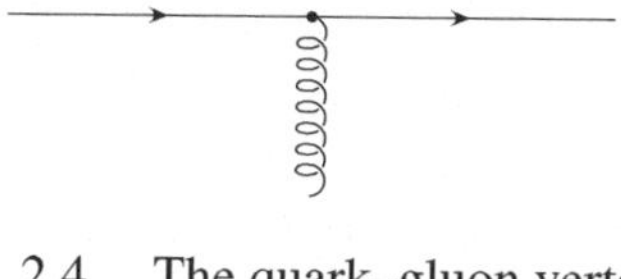

Fig. 2.4. The quark–gluon vertex.

flavor, the effective Lagrangian at leading order in $1/m_Q$ is

$$\mathcal{L}_{\text{eff}} = \sum_{i=1}^{N_h} \bar{Q}_v^{(i)} \, (iv \cdot D) \, Q_v^{(i)}, \tag{2.49}$$

where N_h is the number of heavy quark flavors and all the heavy quarks have the same four-velocity v. The effective Lagrangian in Eq. (2.49) does not depend on the masses or spins of the heavy quarks, and so has a manifest $U(2N_h)$ spin-flavor symmetry under which the $2N_h$ quark fields transform as the fundamental $2N_h$-dimensional representation. There are only $2N_h$ independent components in the N_h fields $Q_v^{(i)}$, because the constraint in Eq. (2.47) eliminates two of the four components in each $Q_v^{(i)}$ spinor field.

2.7 Normalization of states

The standard relativistic normalization for hadronic states is

$$\langle H(p')|H(p)\rangle = 2E_{\mathbf{p}} \, (2\pi)^3 \, \delta^3(\mathbf{p} - \mathbf{p}'), \tag{2.50}$$

where $E_{\mathbf{p}} = \sqrt{|\mathbf{p}|^2 + m_H^2}$. States with the normalization in Eq. (2.50) have mass dimension -1. In HQET, hadron states are labeled by a four-velocity v and a residual momentum k satisfying $v \cdot k = 0$. These states are defined by using the HQET Lagrangian in the $m_Q \to \infty$ limit. They differ from full QCD states by $1/m_Q$ corrections and a normalization factor. The normalization convention in HQET is

$$\langle H(v', k')|H(v, k)\rangle = 2v^0 \, \delta_{vv'} \, (2\pi)^3 \, \delta^3(\mathbf{k} - \mathbf{k}'). \tag{2.51}$$

Possible spin labels are suppressed in Eqs. (2.50) and (2.51). The split between the four-velocity v and the residual momentum is somewhat arbitrary, and the freedom to redefine v by an amount of order Λ_{QCD}/m_Q while changing k by a corresponding amount of order Λ_{QCD} is called reparameterization invariance. We shall explore the consequences of this freedom in Chapter 4. In matrix elements we shall usually take our initial and final hadron states that contain a single heavy quark to have zero residual momentum and not show explicitly the dependence of the state on the residual momentum; i.e., k will be dropped in the labeling of states, $|H(v)\rangle \equiv |H(v, k = 0)\rangle$. The advantage of the normalization in Eq. (2.51) is that it has no dependence on the mass of the heavy quark. A factor of m_H has been removed in comparison with the standard relativistic norm in Eq. (2.50). States normalized by using the HQET convention have mass dimension $-3/2$.

In the remainder of the book, matrix elements in full QCD will be taken between states normalized by using the usual relativistic convention and labeled by the momentum p, whereas matrix elements in HQET will be taken between states normalized by using the HQET convention and labeled by their velocity v. The two normalizations differ by a factor $\sqrt{m_H}$,

$$|H(p)\rangle = \sqrt{m_H}\,[|H(v)\rangle + \mathcal{O}(1/m_Q)]. \tag{2.52}$$

Similarly Dirac spinors $u(p, s)$ labeled by momentum are normalized to satisfy

$$\bar{u}(p, s)\gamma^\mu u(p, s) = 2p^\mu, \tag{2.53}$$

and those labeled by velocity to satisfy

$$\bar{u}(v, s)\gamma^\mu u(v, s) = 2v^\mu. \tag{2.54}$$

The spinors $u(p, s)$ and $u(v, s)$ differ by a factor of $\sqrt{m_H}$

$$u(p, s) = \sqrt{m_H}\, u(v, s). \tag{2.55}$$

2.8 Heavy meson decay constants

Heavy meson decay constants are one of the simplest quantities that can be studied with HQET. The pseudoscalar meson decay constants for the $\bar{B}$ and D mesons are defined by‡

$$\langle 0|\bar{q}\gamma^\mu\gamma_5 Q(0)\,|P(p)\rangle = -if_P\, p^\mu, \tag{2.56}$$

where f_P has mass dimension one. Vector meson decay constants for the D^* and $\bar{B}^*$ mesons are defined by

$$\langle 0|\bar{q}\,\gamma^\mu Q(0)\,|P^*(p, \epsilon)\rangle = f_{P^*}\,\epsilon^\mu, \tag{2.57}$$

where ϵ_μ is the polarization vector of the meson. f_{P^*} has mass dimension two.

The vector and axial currents $\bar{q}\gamma^\mu Q$ and $\bar{q}\gamma^\mu\gamma_5 Q$ can be written in terms of HQET fields,

$$\bar{q}\,\Gamma^\mu Q(0) = \bar{q}\,\Gamma^\mu Q_v(0)\,, \tag{2.58}$$

where $\Gamma^\mu = \gamma^\mu$ or $\gamma^\mu\gamma_5$. There are $\alpha_s(m_Q)$ and $1/m_Q$ corrections to this matching condition, which will be discussed in Chapters 3 and 4, respectively.

The matrix elements required in the heavy quark effective theory are

$$\langle 0|\bar{q}\,\Gamma^\mu Q_v(0)\,|H(v)\rangle, \tag{2.59}$$

where $|H(v)\rangle$ denotes either the P or P^* states with zero residual momentum, normalized using Eq. (2.51). For these matrix elements, it is helpful to reexpress the current $\bar{q}\,\Gamma^\mu Q_v$ in terms of the hadron field $H_v^{(Q)}$ of Eq. (2.19). The current

‡ The pion decay constant f_π defined with the normalization convention in Eq. (2.56) has a value of 131 MeV.

$\bar{q}\,\Gamma^\mu Q_v$ is a Lorentz four vector that transforms as

$$\bar{q}\,\Gamma^\mu Q_v \rightarrow \bar{q}\,\Gamma^\mu D(R)_Q Q_v \tag{2.60}$$

under heavy quark spin transformations, where $D(R)_Q$ is the rotation matrix for a heavy quark field. The representation of the current in terms of $H_v^{(Q)}$ should transform in the same manner as Eq. (2.60) under heavy quark spin transformations. This can be done by using a standard trick: (i) Pretend that Γ^μ transforms as $\Gamma^\mu \rightarrow \Gamma^\mu D(R)_Q^{-1}$ so that the current is an invariant. (ii) Write down operators that are invariant when $Q_v \rightarrow D(R)_Q Q_v$, $\Gamma^\mu \rightarrow \Gamma^\mu D(R)_Q^{-1}$, and $H_v^{(Q)} \rightarrow D(R)_Q H_v^{(Q)}$. (iii) Set Γ^μ to its fixed value γ^μ or $\gamma^\mu\gamma_5$ to obtain the operator with the correct transformation properties.

The current must have a single $H_v^{(Q)}$ field, since the matrix element in Eq. (2.59) contains a single initial-state heavy meson. The field $H_v^{(Q)}$ and Γ^μ can only occur as the product $\Gamma^\mu H_v^{(Q)}$ for the current to be invariant under heavy quark spin symmetry. For Lorentz covariance, the current must have the form

$$\mathrm{Tr}\; X\Gamma^\mu H_v^{(Q)}, \tag{2.61}$$

where X is a Lorentz bispinor. The only parameter that X can depend on is v, so X must have the form $a_0(v^2) + a_1(v^2)\not{v}$, by Lorentz covariance and parity. All dependence on spin has already been included in the indices of the H field, so X can have no dependence on the polarization of the P^* meson. Since $H_v^{(Q)}\not{v} = -H_v^{(Q)}$ and $v^2 = 1$, one can write

$$\bar{q}\Gamma^\mu Q_v = \frac{a}{2}\mathrm{Tr}\;\Gamma^\mu H_v^{(Q)}, \tag{2.62}$$

where $a = [a_0(1) - a_1(1)]$ is an unknown normalization constant that is independent of the mass of the heavy quark Q. Evaluating the trace explicitly gives

$$a \times \begin{cases} -iv^\mu P_v^{(Q)} & \text{if } \Gamma^\mu = \gamma^\mu\gamma_5, \\ P_v^{*(Q)\mu} & \text{if } \Gamma^\mu = \gamma^\mu, \end{cases} \tag{2.63}$$

where $P_v^{(Q)}$ and $P_{v\mu}^{*(Q)}$ are the pseudoscalar and vector fields that destroy the corresponding hadrons. The resulting matrix elements are

$$\begin{aligned} \langle 0|\bar{q}\gamma^\mu\gamma_5 Q_v|P(v)\rangle &= -iav^\mu, \\ \langle 0|\bar{q}\gamma^\mu Q_v|P^*(v)\rangle &= a\epsilon^\mu. \end{aligned} \tag{2.64}$$

Comparing with the definitions of the meson decay constants Eqs. (2.56), (2.57), and using $p^\mu = m_{P^{(*)}} v^\mu$ gives the relations

$$f_P = \frac{a}{\sqrt{m_P}}, \qquad f_{P^*} = a\sqrt{m_{P^*}}. \tag{2.65}$$

Table 2.3. *Heavy meson decay constants from a lattice Monte Carlo simulation*[a]

Decay Constant	Value in MeV
f_D	197 ± 2
f_{D_s}	224 ± 2
f_B	173 ± 4
f_{B_s}	199 ± 3

[a] From the JLQCD Collaboration [S. Aoki et al., Phys. Rev. Lett. 80 (1998), 5711]. Only the statistical errors are quoted.

The factors of $\sqrt{m_P}$ and $\sqrt{m_{P^*}}$ are due to the difference between the normalizations of states in Eqs. (2.50) and (2.51). The P and P^* masses are equal in the heavy quark limit, so one can write the equivalent relations

$$f_P = \frac{a}{\sqrt{m_P}}, \qquad f_{P^*} = m_P f_P, \tag{2.66}$$

which imply that $f_P \propto m_P^{-1/2}$ and $f_{P^*} \propto m_P^{1/2}$. For the D and B system, one finds

$$\frac{f_B}{f_D} = \sqrt{\frac{m_D}{m_B}}, \quad f_{D^*} = m_D f_D, \quad f_{B^*} = m_B f_B. \tag{2.67}$$

The decay constants for the pseudoscalar mesons can be measured by means of the weak leptonic decays $D \to \bar{\ell}\nu_\ell$ and $\bar{B} \to \ell\bar{\nu}_\ell$. The partial width is

$$\Gamma = \frac{G_F^2 |V_{Qq}|^2}{8\pi} f_P^2 m_\ell^2 m_P \left(1 - \frac{m_\ell^2}{m_P^2}\right)^2. \tag{2.68}$$

The only heavy meson decay constant that has been measured is f_{D_s}, from the decays $D_s^+ \to \bar{\mu}\nu_\mu$ and $D_s^+ \to \bar{\tau}\nu_\tau$. However, at the present time, the reported values vary over a large range of ~200–300 MeV. Values of the heavy meson decay constants determined from a lattice Monte Carlo simulation of QCD are shown in Table 2.3. Only statistical errors are quoted. Note that this simulation suggests that there is a substantial correction to the heavy quark symmetry prediction $f_B/f_D = \sqrt{m_D/m_B} \simeq 0.6$.

2.9 $\bar{B} \to D^{(*)}$ form factors

The semileptonic decays of a $\bar{B}$ meson to D and D^* mesons allow one to determine the weak mixing angle V_{cb}. The semileptonic $\bar{B}$ meson decay amplitude is determined by the matrix elements of the weak Hamiltonian:

$$H_W = \frac{4G_F}{\sqrt{2}} V_{cb}[\bar{c}\gamma_\mu P_L b][\bar{e}\gamma^\mu P_L \nu_e]. \tag{2.69}$$

Neglecting higher-order electroweak corrections, the matrix element factors into the product of leptonic and hadronic matrix elements. The hadronic part is the matrix element of the vector or axial vector currents $V^\mu = \bar{c}\gamma^\mu b$ and $A^\mu = \bar{c}\gamma^\mu\gamma_5 b$ between $\bar{B}$ and $D^{(*)}$ states.

It is convenient to write the most general possible matrix element in terms of a few Lorentz invariant amplitudes called form factors. The most general vector current matrix element for $\bar{B} \to D$ must transform as a Lorentz four vector. The only four vectors in the problem are the momenta p and p' of the initial and final mesons, so the matrix element must have the form $ap^\mu + bp'^\mu$. The form factors a and b are Lorentz invariant functions that can only depend on the invariants in the problem, p^2, p'^2 and $p \cdot p'$. Two of the variables are fixed, $p^2 = m_B^2$ and $p'^2 = m_D^2$, and it is conventional to choose $q^2 = (p - p')^2$ as the only independent variable. A similar analysis can be carried out for the other matrix elements. The amplitudes involving the D^* are linear in its polarization vector ϵ and can be simplified by noting that the polarization vector satisfies the constraint $p' \cdot \epsilon = 0$. The conventional choice of form factors allowed by parity and time-reversal invariance is

$$\begin{aligned}
\langle D(p')|V^\mu|\bar{B}(p)\rangle &= f_+(q^2)(p+p')^\mu + f_-(q^2)(p-p')^\mu, \\
\langle D^*(p',\epsilon)|V^\mu|\bar{B}(p)\rangle &= g(q^2)\epsilon^{\mu\nu\alpha\tau}\epsilon^*_\nu(p+p')_\alpha(p-p')_\tau, \\
\langle D^*(p',\epsilon)|A^\mu|\bar{B}(p)\rangle &= -if(q^2)\epsilon^{*\mu} \\
&\quad -i\epsilon^*\cdot p[a_+(q^2)(p+p')^\mu + a_-(q^2)(p-p')^\mu],
\end{aligned} \tag{2.70}$$

where $q = p - p'$, all the form factors are real, and the states have the usual relativistic normalization.

Under parity and time reversal,

$$\begin{aligned}
P|D(p)\rangle &= -|D(p_P)\rangle, & T|D(p)\rangle &= -|D(p_T)\rangle, \\
P|D^*(p,\epsilon)\rangle &= |D^*(p_P,\epsilon_P)\rangle, & T|D^*(p,\epsilon)\rangle &= |D^*(p_T,\epsilon_T)\rangle,
\end{aligned} \tag{2.71}$$

which are the usual transformations for pseudoscalar and vector particles. Here $p = (p^0, \mathbf{p})$, $\epsilon = (\epsilon^0, \boldsymbol{\epsilon})$, and $p_P = p_T = (p^0, -\mathbf{p})$, $\epsilon_P = \epsilon_T = (\epsilon^0, -\boldsymbol{\epsilon})$. Analogous equations hold for the $\bar{B}$ and $\bar{B}^*$. Parity and time-reversal invariance of the strong interactions implies that the matrix elements of currents between two states $|\psi\rangle$ and $|\chi\rangle$ transform as

$$\begin{aligned}
\langle\psi|J^0|\chi\rangle &= \eta_P\langle\psi_P|J^0|\chi_P\rangle, & \langle\psi|J^0|\chi\rangle^* &= \eta_T\langle\psi_T|J^0|\chi_T\rangle, \\
\langle\psi|J^i|\chi\rangle &= -\eta_P\langle\psi_P|J^i|\chi_P\rangle, & \langle\psi|J^i|\chi\rangle^* &= -\eta_T\langle\psi_P|J^i|\chi_P\rangle,
\end{aligned} \tag{2.72}$$

where $\eta_P = 1,\ \eta_T = 1$ if J is the vector current, $\eta_P = -1,\ \eta_T = 1$ if J is the axial current, and $|\chi_P\rangle \equiv P|\chi\rangle$, $|\chi_T\rangle \equiv T|\chi\rangle$, and so on. One can now show that Eq. (2.70) is the most general form factor decomposition. Consider, for example, $\langle D^*(p', \epsilon)|V^\mu|\bar{B}(p)\rangle$. Parity invariance requires that

$$\langle D^*(p', \epsilon)|V^0|\bar{B}(p)\rangle = -\langle D^*(p'_P, \epsilon_P)|V^0|\bar{B}(p_P)\rangle. \qquad (2.73)$$

The only possible tensor combination that changes sign under parity is $\epsilon^{0\nu\alpha\tau}\epsilon^*_\nu p_\alpha\, p'_\tau$, which is proportional to the right-hand side in Eq. (2.70). Time-reversal invariance requires

$$\langle D^*(p', \epsilon)|V^0|\bar{B}(p)\rangle^* = -\langle D^*(p'_T, \epsilon_T)|V^0|\bar{B}(p_T)\rangle, \qquad (2.74)$$

which implies that $g(q^2)$ is real. One can similarly work through the other two cases. The factors of i in Eq. (2.70) depend on the phase convention for the meson states. We have chosen to define the pseudoscalar state to be odd under time reversal. Another choice used is i times this, which corresponds to a state which is even under time reversal. This introduces a factor of i in the last two matrix elements in Eq. (2.70).

It is straightforward to express the differential decay rates $\mathrm{d}\Gamma(\bar{B} \to D^{(*)}e\bar{\nu}_e)/\mathrm{d}q^2$ in terms of the form factors $f_\pm$, f, g, and $a_\pm$. To a very good approximation, the electron mass can be neglected, and consequently a_- and f_- do not contribute to the differential decay rate. For $\bar{B} \to De\bar{\nu}_e$ the invariant decay matrix element is

$$\mathcal{M}(\bar{B} \to De\bar{\nu}_e) = \sqrt{2}G_F V_{cb}\, f_+\,(p+p')^\mu\, \bar{u}(p_e)\gamma_\mu P_L v\big(p_{\nu_e}\big). \qquad (2.75)$$

Squaring and summing over electron spins yields,

$$\begin{aligned}|\mathcal{M}|^2 &= \sum_{\text{spins}} |\mathcal{M}(\bar{B} \to De\bar{\nu}_e)|^2 \\ &= 2G_F^2|V_{cb}|^2|f_+|^2(p+p')^{\mu_1}(p+p')^{\mu_2}\mathrm{Tr}\left[\not{p}_e\gamma_{\mu_1}\not{p}_{\nu_e}\gamma_{\mu_2}P_L\right]. \end{aligned} \qquad (2.76)$$

The differential decay rate is

$$\begin{aligned}\frac{\mathrm{d}\Gamma}{\mathrm{d}q^2}(\bar{B} \to De\bar{\nu}_e) &= \frac{1}{2m_B}\int \frac{d^3p'}{(2\pi)^3 2p'^0}\int \frac{d^3p_e}{(2\pi)^3 2p_e^0} \\ &\times \int \frac{d^3p_{\nu_e}}{(2\pi)^3 2p^0_{\nu_e}}|\mathcal{M}|^2(2\pi)^4\delta^4\big(q - p_e - p_{\nu_e}\big)\delta[q^2 - (p-p')^2], \end{aligned} \qquad (2.77)$$

where q^2 is the hadronic momentum transfer squared, or equivalently, the invariant mass squared of the lepton pair. The integration measure is symmetric with respect to electron and neutrino momenta, so the part of the trace in Eq. (2.76) involving γ_5 does not contribute. It would contribute to the electron spectrum

$\mathrm{d}\Gamma(\bar{B} \to De\bar{\nu}_e)/\mathrm{d}E_e$. The integration over electron and neutrino momenta gives

$$\int \frac{d^3 p_e}{(2\pi)^3 2p_e^0} \int \frac{d^3 p_{\nu_e}}{(2\pi)^3 2p_{\nu_e}^0} \mathrm{Tr}\left[\not{p}_e \gamma_{\mu_1} \not{p}_{\nu_e} \gamma_{\mu_2}\right] (2\pi)^4 \delta^4\left[q - (p_e + p_{\nu_e})\right]$$
$$= \frac{1}{6\pi}\left(q_{\mu_1} q_{\mu_2} - g_{\mu_1\mu_2} q^2\right). \tag{2.78}$$

Finally, using

$$(p + p')^{\mu_1} (p + p')^{\mu_2} \left(q_{\mu_1} q_{\mu_2} - g_{\mu_1\mu_2} q^2\right)$$
$$= \left(q^2 - m_B^2 - m_D^2\right)^2 - 4m_B^2 m_D^2, \tag{2.79}$$

and the two-body phase space formula

$$\int \frac{d^3 p'}{(2\pi)^3 2p'^0} \delta[q^2 - (p - p')^2] = \frac{1}{16\pi^2 m_B^2} \sqrt{\left(q^2 - m_B^2 - m_D^2\right)^2 - 4m_B^2 m_D^2}, \tag{2.80}$$

the differential decay rate in Eq. (2.77) becomes

$$\frac{\mathrm{d}\Gamma}{\mathrm{d}q^2}(\bar{B} \to De\bar{\nu}_e) = \frac{G_F^2 |V_{cb}|^2 |f_+|^2}{192\pi^3 m_B^3} \left[\left(q^2 - m_B^2 - m_D^2\right)^2 - 4m_B^2 m_D^2\right]^{3/2}. \tag{2.81}$$

A similar but more complicated expression holds for $\mathrm{d}\Gamma(\bar{B} \to D^* e\bar{\nu}_e)/\mathrm{d}q^2$.

It is convenient, for comparing with the predictions of HQET, not to write the $\bar{B} \to D^{(*)}$ matrix elements of the vector and axial vector current as in Eq. (2.70), but rather to introduce new form factors that are linear combinations of $f_\pm$, f, g, and $a_\pm$. The four velocities of the $\bar{B}$ and $D^{(*)}$ mesons are $v^\mu = p^\mu/m_B$ and $v'^\mu = p'^\mu/m_{D^{(*)}}$, and the dot product of these four velocities, $w = v \cdot v'$, is related to q^2 by

$$w = v \cdot v' = \left[m_B^2 + m_{D^{(*)}}^2 - q^2\right]/[2m_B m_{D^{(*)}}]. \tag{2.82}$$

The allowed kinematic range for w is

$$0 \le w - 1 \le [m_B - m_{D^{(*)}}]^2/[2m_B m_{D^{(*)}}]. \tag{2.83}$$

The zero-recoil point, at which $D^{(*)}$ is at rest in the $\bar{B}$ rest frame, is $w = 1$. The new form factors $h_\pm$, h_V, and h_{A_j} are expressed as functions of w instead of q^2 and are defined by

$$\begin{aligned}
\frac{\langle D(p')|V^\mu|\bar{B}(p)\rangle}{\sqrt{m_B m_D}} &= h_+(w)\,(v + v')^\mu + h_-(w)\,(v - v')^\mu, \\
\frac{\langle D^*(p', \epsilon)|V^\mu|\bar{B}(p)\rangle}{\sqrt{m_B m_{D^*}}} &= h_V(w) \epsilon^{\mu\nu\alpha\beta} \epsilon_\nu^* v'_\alpha v_\beta, \\
\frac{\langle D^*(p', \epsilon)|A^\mu|\bar{B}(p)\rangle}{\sqrt{m_B m_{D^*}}} &= -ih_{A_1}(w)(w + 1)\epsilon^{*\mu} + ih_{A_2}(w)(\epsilon^* \cdot v)v^\mu \\
&\quad + ih_{A_3}(w)\,(\epsilon^* \cdot v)v'^\mu.
\end{aligned} \tag{2.84}$$

The differential decay rates $\mathrm{d}\Gamma(\bar{B} \to D^{(*)} e \bar{\nu}_e)/\mathrm{d}w$ in terms of these form factors are

$$\frac{\mathrm{d}\Gamma}{\mathrm{d}w}(\bar{B} \to D e \bar{\nu}_e) = \frac{G_F^2 |V_{cb}|^2 m_B^5}{48\pi^3} (w^2-1)^{3/2} r^3 (1+r)^2 \mathcal{F}_D(w)^2,$$

$$\begin{aligned}\frac{\mathrm{d}\Gamma}{\mathrm{d}w}(\bar{B} \to D^* e \bar{\nu}_e) &= \frac{G_F^2 |V_{cb}|^2 m_B^5}{48\pi^3} (w^2-1)^{1/2} (w+1)^2 r^{*3} (1-r^*)^2 \\ &\quad \times \left[1 + \frac{4w}{w+1} \frac{1 - 2wr^* + r^{*2}}{(1-r^*)^2}\right] \mathcal{F}_{D^*}(w)^2, \end{aligned} \tag{2.85}$$

where

$$r = \frac{m_D}{m_B}, \qquad r^* = \frac{m_{D^*}}{m_B}, \tag{2.86}$$

and

$$\begin{aligned}\mathcal{F}_D(w)^2 &= \left[h_+ + \left(\frac{1-r}{1+r}\right) h_-\right]^2, \\ \mathcal{F}_{D^*}(w)^2 &= \left\{2(1 - 2wr^* + r^{*2}) \left[h_{A_1}^2 + \left(\frac{w-1}{w+1}\right) h_V^2\right] \right. \\ &\quad \left. + \left[(1-r^*) h_{A_1} + (w-1)\left(h_{A_1} - h_{A_3} - r^* h_{A_2}\right)\right]^2 \right\} \\ &\quad \times \left\{(1-r^*)^2 + \frac{4w}{w+1}(1 - 2wr^* + r^{*2})\right\}^{-1}. \end{aligned} \tag{2.87}$$

The spin-flavor symmetry of heavy quark effective theory can be used to derive relations between the form factors $h_\pm$, h_V, and h_{A_j}. A transition to the heavy quark effective theory is possible provided the typical momentum transfer to the light degrees of freedom is small compared to the heavy quark masses. In $\bar{B} \to D^{(*)} e \bar{\nu}_e$ semileptonic decay, q^2 is not small compared with $m_{c,b}^2$. However, this variable does not determine the typical momentum transfer to the light degrees of freedom. A rough measure of that is the momentum transfer that must be given to the light degrees of freedom so that they recoil with the $D^{(*)}$. The light degrees of freedom in the initial and final hadrons have momentum of order $\Lambda_{\mathrm{QCD}} v$ and $\Lambda_{\mathrm{QCD}} v'$, respectively, since their velocity is fixed to be the same as the heavy quark velocity. The momentum transfer for the light system is then $q_{\text{light}}^2 \sim (\Lambda_{\mathrm{QCD}} v - \Lambda_{\mathrm{QCD}} v')^2 = 2\Lambda_{\mathrm{QCD}}^2 (1-w)$. Heavy quark symmetry should hold, provided

$$2\Lambda_{\mathrm{QCD}}^2 (w-1) \ll m_{b,c}^2. \tag{2.88}$$

The heavy meson form factors are expected to vary on the scale $q_{\text{light}}^2 \sim \Lambda_{\mathrm{QCD}}^2$, i.e., on the scale $w \sim 1$.

The six form factors can be computed in terms of a single function using heavy quark symmetry. The QCD matrix elements required are of the form $\langle H^{(c)}(p')|\bar{c}\,\Gamma b|H^{(b)}(p)\rangle$, where $\Gamma=\gamma^\mu$, $\gamma^\mu\gamma_5$ and $H^{(Q)}$ is either $P^{(Q)}$ or $P^{*(Q)}$. At leading order in $1/m_{c,b}$ and $\alpha_s(m_{c,b})$, the current $\bar{c}\,\Gamma b$ can be replaced by the current $\bar{c}_{v'}\Gamma b_v$ involving heavy quark fields and the heavy mesons states $|H^{(Q)}(p^{(\prime)})\rangle$ by the corresponding ones in HQET $|H^{(Q)}(v^{(\prime)})\rangle$. One can then use a trick similar to that used for the meson decay constants: the current is invariant under spin transformations on the $c_{v'}$ and b_v quark fields, provided that Γ transforms as $D(R)_c\Gamma D(R)_b^{-1}$ where $D(R)_c$ and $D(R)_b$ are the heavy quark spin rotation matrices for c and b quarks, respectively. For the required matrix elements one represents the current by operators that contain one factor each of $\bar{H}_{v'}^{(c)}$ and $H_v^{(b)}$, so that a meson containing a b quark is converted to one containing a c quark. Invariance under the b and c quark spin symmetries requires that the operators should be of the form $\bar{H}_{v'}^{(c)}\Gamma H_v^{(b)}$, so that the factors of $D(R)_{b,c}$ cancel between the Γ matrix and the H fields. Lorentz covariance then requires that

$$\bar{c}_{v'}\Gamma b_v = \mathrm{Tr}\, X\bar{H}_{v'}^{(c)}\Gamma H_v^{(b)}, \tag{2.89}$$

where X is the most general possible bispinor that one can construct using the available variables, v and v'. The most general form for X with the correct parity and time-reversal properties is

$$X = X_0 + X_1\not{v} + X_2\not{v}' + X_3\not{v}\not{v}', \tag{2.90}$$

where the coefficients are functions of $w=v\cdot v'$. Other allowed terms can all be written as linear combinations of the X_i. For example, $\not{v}'\not{v}=2w-\not{v}\not{v}'$, and so on. The relations $\not{v}H_v^{(b)}=H_v^{(b)}$ and $\not{v}'\bar{H}_{v'}^{(c)}=-\bar{H}_{v'}^{(c)}$ imply that all the terms in Eq. (2.90) are proportional to the first, so one can write

$$\bar{c}_{v'}\Gamma\, b_v = -\xi(w)\mathrm{Tr}\,\bar{H}_{v'}^{(c)}\,\Gamma H_v^{(b)}, \tag{2.91}$$

where the coefficient is conventionally written as $-\xi(w)$. Evaluating the trace in Eq. (2.91) gives the required HQET matrix elements

$$\begin{aligned}\langle D(v')|\bar{c}_{v'}\,\gamma_\mu\, b_v|\bar{B}(v)\rangle &= \xi(w)\,[v_\mu + v'_\mu],\\ \langle D^*(v',\epsilon)|\bar{c}_{v'}\gamma_\mu\gamma_5 b_v|\bar{B}(v)\rangle &= -i\xi(w)\,[(1+w)\epsilon^*_\mu - (\epsilon^*\cdot v)v'_\mu],\\ \langle D^*(v',\epsilon)|\bar{c}_{v'}\gamma_\mu b_v|\bar{B}(v)\rangle &= \xi(w)\,\epsilon_{\mu\nu\alpha\beta}\epsilon^{*\nu}v'^\alpha v^\beta.\end{aligned} \tag{2.92}$$

Equations (2.92) are the implications of heavy quark spin symmetry for the $\bar{B}\to D^{(*)}$ matrix elements of the axial vector and vector currents. The function

$\xi(w)$ is independent of the charm and bottom quark masses. Heavy quark flavor symmetry implies the normalization condition

$$\xi(1) = 1. \tag{2.93}$$

To derive this result, consider the forward matrix element of the vector current $\bar{b}\gamma^\mu b$ between $\bar{B}$ meson states. To leading order in $1/m_b$, the operator $\bar{b}\gamma^\mu b$ can be replaced by $\bar{b}_v\gamma_\mu b_v$. The forward matrix element can then be obtained from Eq. (2.92) by setting $v' = v$, and letting $c \to b$, $D \to \bar{B}$,

$$\frac{\langle \bar{B}(p)|\bar{b}\gamma_\mu b|\bar{B}(p)\rangle}{m_B} = \langle \bar{B}(v)|\bar{b}_v\gamma_\mu b_v|\bar{B}(v)\rangle = 2\,\xi(w=1)\,v_\mu. \tag{2.94}$$

Note that $\xi(w)$ is independent of the quark masses, and so has the same value in Eqs. (2.92) and (2.94). Equivalently, heavy quark flavor symmetry allows one to replace D by $\bar{B}$ in Eq. (2.92). The left-hand side of Eq. (2.94) with $\mu = 0$ is the matrix element of b-quark number between $\bar{B}$ mesons, and so has the value $2v_0$. This implies that $\xi(1) = 1$.

Functions of $w = v \cdot v'$ like ξ occur often in the analysis of matrix elements and are called Isgur-Wise functions. Eq. (2.92) predicts relations between the form factors in Eqs. (2.84):

$$\begin{aligned} h_+(w) = h_V(w) = h_{A_1}(w) = h_{A_3}(w) &= \xi(w), \\ h_-(w) = h_{A_2}(w) &= 0. \end{aligned} \tag{2.95}$$

This equation implies that

$$\mathcal{F}_D(w) = \mathcal{F}_{D^*}(w) = \xi(w). \tag{2.96}$$

There is experimental support for the utility of the $m_{c,b} \to \infty$ limit for describing $\bar{B} \to D^{(*)}e\bar{\nu}_e$ decays. Figure 2.5 shows a plot of the ratio $\mathcal{F}_{D^*}(w)/\mathcal{F}_D(w)$ as a

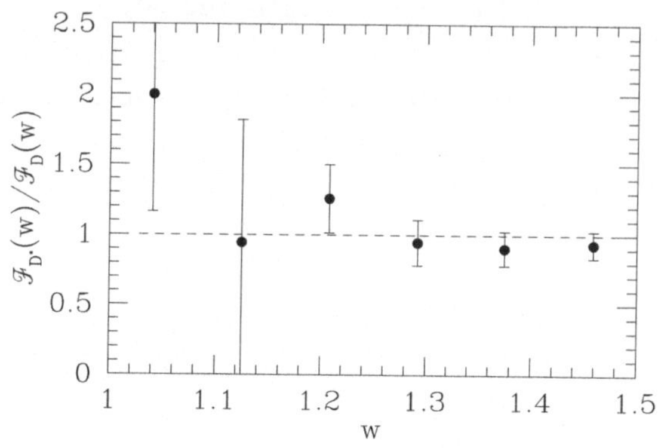

Fig. 2.5. The measured ratio $\mathcal{F}_{D^*}(w)/\mathcal{F}_D(w)$ as a function of w. The data are from the ALEPH Collaboration [D. Buskulic et al., Phys. Lett. B395 (1997) 373].

function of w using data from the ALEPH collaboration. It shows that $\mathcal{F}_{D^*}(w)$ is indeed near $\mathcal{F}_D(w)$. Note that the experimental errors become large as w approaches unity. This is partly because the differential rates $d\Gamma/dw$ vanish at $w = 1$. In addition to comparing the D and D^* decay rates, there is experimental information on the individual form factors in $\bar{B} \to D^* e \bar{\nu}_e$ decay. It is convenient to define two ratios of these form factors:

$$R_1 = \frac{h_V}{h_{A_1}}, \quad R_2 = \frac{h_{A_3} + r h_{A_2}}{h_{A_1}}. \tag{2.97}$$

In the $m_{c,b} \to \infty$ limit, heavy quark spin symmetry implies that $R_1 = R_2 = 1$. Assuming the form factors h_j have the same shape in w, the CLEO collaboration has obtained the experimental values [J. E. Duboscq et al., Phys. Rev. Lett. 76 (1996) 3898]

$$R_1 = 1.18 \pm 0.3, \quad R_2 = 0.71 \pm 0.2. \tag{2.98}$$

There is a simple physical reason why a single Isgur-Wise function is needed for the matrix elements in Eq. (2.92). In the $m_{c,b} \to \infty$ limit, the spin of the light degrees of freedom is a good quantum number. Since $\bar{c}_{v'} \Gamma b_v$ does not act on the light degrees of freedom, their helicity, h_ℓ, is conserved in the transitions it mediates. For $\bar{B} \to D^{(*)}$ matrix elements, there are two helicity amplitudes corresponding to $h_\ell = 1/2$ and $h_\ell = -1/2$. However, they must be equal by parity invariance and therefore there is only one Isgur-Wise function. There are cases when more than one Isgur-Wise function occurs. For example, in $\Omega_b \to \Omega_c^{(*)} e \bar{\nu}_e$ decay, the initial and final hadrons have $s_\ell = 1$. Thus there are two independent helicity amplitudes $h_\ell = 0$ and $h_\ell = \pm 1$, and consequently, two Isgur-Wise functions occur (see Problem 10).

2.10 $\Lambda_c \to \Lambda$ form factors

Another interesting application of heavy quark symmetry is to the weak decay $\Lambda_c \to \Lambda \bar{e} \nu_e$. This decay is an example of a heavy $\to$ light transition, in which a heavy quark decays to a light quark. The most general weak decay form factors can be written in the form

$$\begin{aligned}
\langle \Lambda(p', s') | \bar{s} \gamma^\mu c | \Lambda_c(p, s) \rangle &= \bar{u}(p', s') [f_1 \gamma^\mu + i f_2 \sigma^{\mu\nu} q_\nu + f_3 q^\mu] u(p, s), \\
\langle \Lambda(p', s') | \bar{s} \gamma^\mu \gamma_5 c | \Lambda_c(p, s) \rangle &= \bar{u}(p', s') [g_1 \gamma^\mu + i g_2 \sigma^{\mu\nu} q_\nu + g_3 q^\mu] \gamma_5 u(p, s),
\end{aligned} \tag{2.99}$$

where $q = p - p'$ and $\sigma_{\mu\nu} = i[\gamma_\mu, \gamma_\nu]/2$. The form factors f_i and g_i are functions of q^2. Heavy quark spin symmetry on the c-quark constrains the general form factor decomposition in Eq. (2.99). Making the transition to HQET, one can write the left-hand side of Eq. (2.99) as

$$\langle \Lambda(p', s') | \bar{s}\, \Gamma c_v | \Lambda_c(v, s) \rangle, \tag{2.100}$$

where $\bar{s}\Gamma c \to \bar{s}\Gamma c_v$ at leading order in $1/m_c$. The matrix element in Eq. (2.100) has the same form factor expansion as Eq. (2.99) with $u(p, s) \to u(v, s)$. The $\sqrt{m_{\Lambda_c}}$ difference between Eqs. (2.99) and (2.100) in the normalization of states is compensated by the same factor in the normalization of spinors. The most general form for the matrix element in Eq. (2.100) consistent with spin symmetry on the c quark is

$$\langle \Lambda(p', s') | \bar{s} \Gamma c_v | \Lambda_c(v, s) \rangle = \bar{u}(p', s') X \Gamma u(v, s), \tag{2.101}$$

where X is the most general bispinor that can be constructed out of p' and v. Note that s and s' cannot be used, because the fermion spin is encoded in the matrix indices of the spinors. The decomposition of X is

$$X = F_1 + F_2 \not{v}, \tag{2.102}$$

where F_i are functions of $v \cdot p'$, and we have used the constraints

$$\not{v} u(v, s) = u(v, s), \qquad \not{p}' u(p', s') = m_\Lambda u(p', s') \tag{2.103}$$

to reduce the number of independent terms. Substituting Eq. (2.102) into Eq. (2.101) and comparing with Eq. (2.99) gives the relations

$$\begin{aligned} f_1 = g_1 &= F_1 + \frac{m_\Lambda}{m_{\Lambda_c}} F_2, \\ f_2 = f_3 = g_2 = g_3 &= \frac{1}{m_{\Lambda_c}} F_2, \end{aligned} \tag{2.104}$$

so that the six form factors f_i, g_i can be written in terms of two functions $F_{1,2}$. The heavy $\to$ light form factors $F_{1,2}$ are expected to vary on the scale $v \cdot p' \sim \Lambda_{\rm QCD}$.

These relations between form factors have implications for the polarization of the Λ's produced in Λ_c decay. Equation (2.104) implies that in the $m_c \to \infty$ limit, the polarization variable

$$\alpha = - \left. \frac{2 f_1 g_1}{f_1^2 + g_1^2} \right|_{q^2=0} \tag{2.105}$$

is equal to -1. The CLEO Collaboration [G. Crawford et al., Phys. Rev. Lett. 75 (1995) 624] finds that, averaged over all q^2, $\alpha = -0.82 \pm 0.10$ consistent with expectations based on charm quark spin symmetry.

2.11 $\Lambda_b \to \Lambda_c$ form factors

The semileptonic weak decay $\Lambda_b \to \Lambda_c e\bar{\nu}_e$ form factors are even more constrained by heavy quark symmetry than the $\Lambda_c \to \Lambda\bar{e}\nu_e$ form factors discussed above, because one can use heavy quark symmetry on both the initial and final baryons. The most general weak decay form factors for $\Lambda_b \to \Lambda_c$ decay are conventionally written as

$$\begin{aligned}\langle\Lambda_c(p',s')|\bar{c}\gamma^\mu b|\Lambda_b(p,s)\rangle &= \bar{u}(p',s')[f_1\gamma^\mu + f_2 v^\mu + f_3 v'^\mu]u(p,s),\\ \langle\Lambda_c(p',s')|\bar{c}\gamma^\mu\gamma_5 b|\Lambda_b(p,s)\rangle &= \bar{u}(p',s')[g_1\gamma^\mu + g_2 v^\mu + g_3 v'^\mu]\gamma_5 u(p,s),\end{aligned} \tag{2.106}$$

where f_i and g_i are functions of w. We have taken the general decomposition from Eq. (2.99) and rewritten q^μ and $\sigma^{\mu\nu}q_\nu$ in terms of γ^μ, v^μ and v'^μ. Making the transition to HQET, the matrix element

$$\langle\Lambda_c(v',s')|\bar{c}_{v'}\,\Gamma\, b_v|\Lambda_b(v,s)\rangle = \zeta(w)\bar{u}(v',s')\Gamma u(v,s) \tag{2.107}$$

by heavy quark spin symmetry on the b and c quark fields. Thus we obtain

$$f_1(w) = g_1(w) = \zeta(w), \qquad f_2 = f_3 = g_2 = g_3 = 0. \tag{2.108}$$

The six form factors can be written in terms of the single Isgur-Wise function $\zeta(w)$. As in the meson case

$$\zeta(1) = 1, \tag{2.109}$$

since the form factor of $\bar{b}\gamma^\mu b$ for $\Lambda_b \to \Lambda_b$ transitions at $w=1$ is b-quark number. The heavy $\to$ heavy relations in Eq. (2.108) are a special case of the heavy $\to$ light relations in Eq. (2.104), with the additional restrictions $F_2=0$ and $F_1(v\cdot v'=1)=1$.

2.12 Problems

1. In the $m_Q \to \infty$ limit, show that the propagator for a heavy antiquark with momentum $p_{\bar{Q}} = m_Q v + k$ is

$$\frac{i}{v\cdot k + i\varepsilon}\left(\frac{1-\not{v}}{2}\right),$$

while the heavy antiquark–gluon vertex is

$$ig(T^A)^T v_\mu.$$

2. Compare the theoretical expectation for the ratio $\Gamma(D_1 \to D^*\pi)/\Gamma(D_2^* \to D^*\pi)$ with its experimental value. Discuss your result.

3. Consider the following heavy-light matrix elements of the vector and axial vector currents

$$\langle V(p',\epsilon)|\bar{q}\gamma_\mu\gamma_5 Q|P^{(Q)}(p)\rangle = -if^{(Q)}\epsilon^*_\mu - i\epsilon^*\cdot p\big[a_+^{(Q)}(p+p')_\mu + a_-^{(Q)}(p-p')_\mu\big],$$
$$\langle V(p',\epsilon)|\bar{q}\gamma_\mu Q|P^{(Q)}(p)\rangle = g^{(Q)}\epsilon_{\mu\nu\lambda\sigma}\epsilon^{*\nu}(p+p')^\lambda(p-p')^\sigma,$$

where $p=m_{P^{(Q)}}v$. The form factors $f^{(Q)}$, $a_\pm^{(Q)}$ and $g^{(Q)}$ are functions of $y=v\cdot p'$. V is a low-lying vector meson, i.e., either a ρ or K^* depending on the light quark flavor quantum numbers of q and $P^{(Q)}$. Show that in the $m_{b,c}\to\infty$ limit

$$f^{(b)}(y) = (m_b/m_c)^{1/2} f^{(c)}(y),$$
$$g^{(b)}(y) = (m_c/m_b)^{1/2} g^{(c)}(y),$$
$$a_+^{(b)}(y) + a_-^{(b)}(y) = (m_c/m_b)^{3/2}\big[a_+^{(c)}(y) + a_-^{(c)}(y)\big],$$
$$a_+^{(b)}(y) - a_-^{(b)}(y) = (m_c/m_b)^{1/2}\big[a_+^{(c)}(y) - a_-^{(c)}(y)\big].$$

Discuss how these results may be used to determine V_{ub} from data on the semileptonic decays $B\to\rho e\bar{\nu}_e$ and $D\to\rho\bar{e}\nu_e$.

4. Consider the matrix element

$$\langle V(p',\epsilon)|\bar{q}\sigma_{\mu\nu} Q|P^{(Q)}(p)\rangle = -ig_+^{(Q)}\epsilon_{\mu\nu\lambda\sigma}\epsilon^{*\lambda}(p+p')^\sigma - ig_-^{(Q)}\epsilon_{\mu\nu\lambda\sigma}\epsilon^{*\lambda}(p-p')^\sigma$$
$$-ih^{(Q)}\epsilon_{\mu\nu\lambda\sigma}(p+p')^\lambda(p-p')^\sigma(\epsilon^*\cdot p).$$

Show that in the $m_Q\to\infty$ limit the form factors $g_\pm^{(Q)}$ and $h^{(Q)}$ are related to those in Problem 3 by

$$g_+^{(Q)} - g_-^{(Q)} = -m_Q g^{(Q)},$$
$$g_+^{(Q)} + g_-^{(Q)} = f^{(Q)}/2m_Q + \frac{p\cdot p'}{m_Q}g^{(Q)},$$
$$h^{(Q)} = -\frac{g^{(Q)}}{m_Q} + \frac{a_+^{(Q)} - a_-^{(Q)}}{2m_Q}.$$

5. Verify the expressions for the $P\to\ell\bar{\nu}_e$, $\bar{B}\to De\bar{\nu}_e$, and $\bar{B}\to D^*e\bar{\nu}_e$ decay rates given in the text.

6. The fields $D_2^{*\mu\nu}$ and D_1^μ destroy the spin-two and spin-one members of the excited doublet of charmed mesons with $s_\ell = 3/2$ and positive parity. Show that

$$F_v^\mu = \frac{(1+\not{v})}{2}\left\{D_2^{*\mu\nu}\gamma_\nu - \sqrt{\frac{3}{2}}D_1^\nu\gamma_5\left[g_\nu^\mu - \frac{1}{3}\gamma_\nu(\gamma^\mu - v^\mu)\right]\right\},$$

satisfies

$$\not{v}F_v^\mu = F_v^\mu, \qquad F_v^\mu\not{v} = -F_v^\mu, \qquad F_v^\mu\gamma_\mu = F_v^\mu v_\mu = 0,$$

and that under heavy charm quark spin transformations

$$F_v^\mu \to D(R)_c F_v^\mu.$$

7. Use Lorentz, parity, and time-reversal invariance to argue that the form factor decompositions of matrix elements of the weak vector and axial vector $b \to c$ currents are

$$\frac{\langle D_1(p',\epsilon)|V^\mu|\bar{B}(p)\rangle}{\sqrt{m_B m_{D_1}}} = -i f_{V_1}\epsilon^{*\mu} - i\big(f_{V_2} v^\mu + f_{V_3} v'^\mu\big)(\epsilon^* \cdot v),$$

$$\frac{\langle D_1(p',\epsilon)|A^\mu|\bar{B}(p)\rangle}{\sqrt{m_B m_{D_1}}} = f_A \epsilon^{\mu\alpha\beta\gamma}\epsilon^*_\alpha v_\beta v'_\gamma,$$

$$\frac{\langle D_2^*(p',\epsilon)|A^\mu|\bar{B}(p)\rangle}{\sqrt{m_B m_{D_2^*}}} = -i k_{A_1}\epsilon^{*\mu\alpha} v_\alpha + \big(k_{A_2} v^\mu + k_{A_3} v'^\mu\big)\epsilon^*_{\alpha\beta} v^\alpha v^\beta,$$

$$\frac{\langle D_2^*(p',\epsilon)|V^\mu|\bar{B}(p)\rangle}{\sqrt{m_B m_{D_2^*}}} = k_V \epsilon^{\mu\alpha\beta\gamma}\epsilon^*_{\alpha\sigma} v^\sigma v_\beta v'_\gamma,$$

where v' is the four velocity of the final charmed meson and v the four velocity of the $\bar{B}$ meson. Note that the D_1 polarization vector is denoted by ϵ_α while the D_2^* polarization tensor is denoted by $\epsilon_{\alpha\beta}$.

8. Show that

$$\frac{d\Gamma}{dw}(\bar{B} \to D_1 e\bar{\nu}_e) = \frac{G_F^2|V_{cb}|^2 m_B^5}{48\pi^3} r_1^3\sqrt{w^2-1}\{\big[(w-r_1)f_{V_1} + (w^2-1)\big(f_{V_3} + r_1 f_{V_2}\big)\big]^2 + 2\big(1 - 2r_1 w + r_1^2\big)\big[f_{V_1}^2 + (w^2-1)f_A^2\big]\},$$

$$\frac{d\Gamma}{dw}(\bar{B} \to D_2^* e\bar{\nu}_e) = \frac{G_F^2|V_{cb}|^2 m_B^5}{48\pi^3} r_2^3 (w^2-1)^{3/2}\left\{\frac{2}{3}\big[(w-r_2)k_{A_1} + (w^2-1)\big(k_{A_3} + r_2 k_{A_2}\big)\big]^2 + \big[1 - 2r_2 w + r_2^2\big]\big[k_{A_1}^2 + (w^2-1)k_V^2\big]\right\},$$

where the form factors, which are functions of $w = v\,.\,v'$, are defined in problem 7.

9. Argue that for $B \to D_1$ and $B \to D_2^*$ matrix elements, heavy quark spin symmetry implies that one can use

$$\bar{c}_{v'}\Gamma b_v = \tau(w)\mathrm{Tr}\{v_\sigma \bar{F}_{v'}^\sigma \Gamma H_v^{(b)}\},$$

where $\tau(w)$ is a function of w, and F_v^μ was defined in Problem 6. Deduce the following expressions for the form factors

$$\begin{aligned} \sqrt{6}\, f_A &= -(w+1)\tau, & k_V &= -\tau, \\ \sqrt{6}\, f_{V_1} &= -(1-w^2)\tau, & k_{A_1} &= -(1+w)\tau, \\ \sqrt{6}\, f_{V_2} &= -3\tau, & k_{A_2} &= 0, \\ \sqrt{6}\, f_{V_3} &= (w-2)\tau, & k_{A_3} &= 0. \end{aligned}$$

Only the form factor f_{V_1} can contribute to the weak matrix elements at zero recoil, $w = 1$. Notice that $f_{V_1}(1) = 0$ for any value of $\tau(1)$. Is there a normalization condition on $\tau(1)$ from heavy quark flavor symmetry?

10. The ground-state baryons with two strange quarks and a heavy quark decay weakly, $\Omega_b \to \Omega_c^{(*)} e\bar{\nu}_e$. They occur in a $s_\ell = 1$ doublet, and the spin-$1/2$ and spin-$3/2$ members are denoted

by Ω_Q and Ω_Q^* respectively. Show that the field

$$S_{v\mu}^{(Q)} = \left[\frac{1}{\sqrt{3}}(\gamma_\mu + v_\mu)\gamma_5 \Omega_v^{(Q)} + \Omega_{v\mu}^{*(Q)}\right]$$

transforms under heavy quark spin symmetry as

$$S_{v\mu}^{(Q)} \to D(R)_Q S_{v\mu}^{(Q)}.$$

Here $\Omega_v^{(Q)}$ is a spin-1/2 field that destroys a Ω_Q state with amplitude $u(v, s)$ and $\Omega_{v\mu}^{*(Q)}$ is a spin-3/2 field that destroys a Ω_Q^* state with amplitude $u_\mu(v, s)$. Here $u_\mu(v, s)$ is a Rarita-Schwinger spinor that satisfies $\not{v} u_\mu(v, s) = u_\mu(v, s)$, $v^\mu u_\mu(v, s) = \gamma^\mu u_\mu(v, s) = 0$. Argue that for $\Omega_Q \to \Omega_Q^{(*)}$ matrix elements heavy quark symmetry implies that

$$\bar{c}_{v'} \Gamma b_v = \mathrm{Tr}\bar{S}_{v'\mu}^{(c)} \Gamma S_{vv}^{(b)} [-g^{\mu v}\lambda_1(w) + v^\mu v'^v \lambda_2(w)].$$

Show that heavy quark flavor symmetry requires the normalization condition

$$\lambda_1(1) = 1$$

at zero recoil.

2.13 References

Semileptonic $\bar{B} \to D^{(*)}$ at zero recoil was studied in:

S. Nussinov and W. Wetzel, Phys. Rev. D36 (1987) 130
M.B. Voloshin and M.A. Shifman, Sov. J. Nucl. Phys. 47 (1988) 511 [Yad. Fiz. 47 (1988) 801]
These papers use the physics behind heavy quark symmetry to deduce the matrix elements at zero recoil.

Early uses of heavy quark symmetry appear in:

N. Isgur and M.B. Wise, Phys. Lett. B232 (1989) 113, B237 (1990) 527

Implications of heavy quark symmetry for spectroscopy was studied in:

N. Isgur and M.B. Wise, Phys. Rev. Lett. 66 (1991) 1130

For spectroscopy of the excited heavy mesons using the nonrelativistic constituent quark model, see:

J. Rosner, Comm. Nucl. Part. Phys. 16 (1986) 109
E. Eichten, C.T. Hill, and C. Quigg, Phys. Rev. Lett. 71 (1993) 4116

The effective field theory was formulated in:

B. Grinstein, Nucl. Phys. B339 (1990) 253
E. Eichten and B. Hill, Phys. Lett. B234 (1990) 511
H. Georgi, Phys. Lett. B240 (1990) 447

see also:

J.G. Korner and G. Thompson, Phys. Lett. B264 (1991) 185
T. Mannel, W. Roberts, and Z. Ryzak, Nucl. Phys. B368 (1992) 204

The H field formalism was introduced in:

J.D. Bjorken, Invited talk at Les Rencontre de la Valle d'Aoste, La Thuile, Italy, 1990
A.F. Falk, H. Georgi, B. Grinstein, and M.B. Wise, Nucl. Phys. B343 (1990) 1

and was extended to hadrons with arbitrary spin in:

A.F. Falk, Nucl. Phys. B378 (1992) 79

For counting the number of independent amplitudes using the helicity formalism, see:

H.D. Politzer, Phys. Lett. B250 (1990) 128

Fragmentation to heavy hadrons was discussed in:

A.F. Falk and M.E. Peskin, Phys. Rev. D49 (1994) 3320
R.L. Jaffe and L. Randall, Nucl. Phys. B412 (1994) 79

Baryon decays were studied in:

T. Mannel, W. Roberts, and Z. Ryzak, Nucl. Phys. B355 (1991) 38
F. Hussain, J.G. Korner, M. Kramer, and G. Thompson, Z. Phys. C51 (1991) 321
F. Hussain, D.-S. Liu, M. Kramer, and J.G. Korner, Nucl. Phys. B370 (1992) 2596
N. Isgur and M.B. Wise, Nucl. Phys. B348 (1991) 276
H. Georgi, Nucl. Phys. B348 (1991) 293

Heavy quark symmetry relations for heavy-light form factors were discussed in:

N. Isgur and M.B. Wise, Phys. Rev. D42 (1990) 2388

$\bar{B}$ semileptonic decay to excited charm mesons was considered in:

N. Isgur and M.B. Wise, Phys. Rev. D43 (1991) 819

3
Radiative corrections

The previous chapter derived some simple consequences of heavy quark symmetry ignoring $1/m_Q$ and radiative corrections. This chapter discusses how radiative corrections can be systematically included in HQET computations. The two main issues are the computation of radiative corrections in the matching between QCD and HQET, and the renormalization of operators in the effective theory. The renormalization of the effective theory is considered first, because it is necessary to understand this before computing corrections to the matching conditions. The $1/m_Q$ corrections will be discussed in the next chapter.

3.1 Renormalization in HQET

The fields and the coupling in the HQET Lagrange density Eq. (2.49) are actually bare quantities,

$$\mathcal{L}_{\text{eff}} = i\bar{Q}_v^{(0)} v^\mu \left[\partial_\mu + ig^{(0)} A_\mu^{(0)}\right] Q_v^{(0)}, \tag{3.1}$$

where the superscript (0) denotes bare quantities. It is convenient to define renormalized fields that have finite Green's functions. The renormalized heavy quark field is related to the bare one by wave-function renormalization,

$$Q_v = \frac{1}{\sqrt{Z_h}}\, Q_v^{(0)}. \tag{3.2}$$

The coupling constant $g^{(0)}$ and the gauge field $A_\mu^{(0)}$ are also related to the renormalized coupling and gauge field by multiplicative renormalization. In the background field gauge, gA_μ is not renormalized, so $g^{(0)} A_\mu^{(0)} = g\mu^{\epsilon/2} A_\mu$, where $n = 4 - \epsilon$ is the dimension of space–time.

In terms of renormalized quantities, the HQET Lagrangian becomes

$$\begin{aligned}\mathcal{L}_{\text{eff}} &= iZ_h\, \bar{Q}_v v^\mu \left(\partial_\mu + ig\mu^{\epsilon/2} A_\mu\right) Q_v \\ &= i\bar{Q}_v v^\mu \left(\partial_\mu + ig\mu^{\epsilon/2} A_\mu\right) Q_v + \text{ counterterms.}\end{aligned} \tag{3.3}$$

Fig. 3.1. Heavy quark loop graph, which vanishes in the effective theory. Heavy quark propagators are denoted by a double line.

Equation (3.3) has been written in $n = 4-\epsilon$ dimensions, with μ the dimensionful scale parameter of dimensional regularization.

Heavy quarks do not effect the renormalization constants for light quark fields Z_q, the gluon field Z_A, and the strong coupling Z_g, because heavy quark loops vanish in the effective theory. That loops do not occur is evident from the propagator in Eq. (2.41). In the rest frame $v = v_r$ the propagator $i/(k \cdot v + i\varepsilon)$ has one pole below the real axis at $k^0 = -i\varepsilon$. A closed heavy quark loop graph such as in Fig. 3.1 involves an integration over the loop momentum k. The heavy quark propagators in the loop both have poles below the real axis, so the k^0 integral can be closed in the upper half-plane, giving zero for the loop integral. The HQET field Q_v annihilates a heavy quark but does not create the corresponding antiquark.

In the full theory of QCD, the light quark wave-function renormalization Z_q is independent of the quark mass in the $\overline{\text{MS}}$ scheme. A heavy quark with mass m_Q contributes to the QCD β function even for $\mu \ll m_Q$. At first glance, this would imply that heavy particle effects do not decouple at low energies. This nondecoupling is an artifact of the $\overline{\text{MS}}$ scheme. The finite parts of loop graphs have a logarithmic dependence on the quark mass and become large as $\mu \ll m_Q$. One can show that the logarithmic dependence of the finite parts exactly cancels the logarithmic heavy quark contribution to the renormalization group equation, so that the total heavy quark contribution vanishes as $\mu \ll m_Q$. This cancellation can be made manifest in the zero heavy quark sector by constructing an effective theory for $\mu < m_Q$ in which the heavy quark has been integrated out. Such effective theories were considered in Sec. 1.5 of Chapter 1. Similarly, in HQET, one matches at $\mu = m_Q$ to a new theory in which the Dirac propagator for the heavy quark is replaced by the HQET propagator Eq. (2.41). This changes the renormalization scheme for the heavy quarks, so that Z_h for the heavy quark differs from Z_q for the light quarks.

Z_h can be computed by studying the one-loop correction to the heavy quark propagator in Fig. 3.2. In the Feynman gauge, the graph is

$$\int \frac{d^n q}{(2\pi)^n} \left(- igT^A \mu^{\epsilon/2} \right) v_\lambda \frac{i}{(q+p)\cdot v} \left(- igT^A \mu^{\epsilon/2} \right) v^\lambda \frac{(-i)}{q^2}$$
$$= -\left(\frac{4}{3}\right) g^2 \mu^\epsilon \int \frac{d^n q}{(2\pi)^n} \frac{1}{q^2 \, v\cdot(q+p)}, \tag{3.4}$$

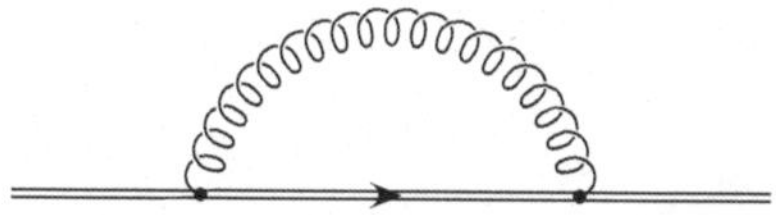

Fig. 3.2. Gluon interaction with a heavy quark.

where p is the external residual momentum, q is the loop momentum, and we have used the identity $T^A T^A = (4/3)\mathbb{1}$ for the **3** of $SU(3)$. The one-loop wave-function renormalization is given by the ultraviolet divergent part of Eq. (3.4). If one expands in $v \cdot p$, Eq. (3.4) is also infrared divergent, and it is convenient to regulate the infrared divergence by giving the gluon a mass m that will be set to zero at the end of the computation. This infrared regulator allows one to isolate the ultraviolet divergence by computing the $1/\epsilon$ term in the integral. The regulated integral that has to be evaluated is

$$-\left(\frac{4}{3}\right) g^2 \mu^\epsilon \int \frac{d^n q}{(2\pi)^n} \frac{1}{(q^2 - m^2)[v \cdot (q+p)]}, \tag{3.5}$$

where m is the gluon mass. The integral Eq. (3.5) will be computed in detail, since it provides an example of some standard tricks that are useful in computing loop graphs in HQET. The denominators can be combined by using the identity

$$\frac{1}{a^r b^s} = 2^s \frac{\Gamma(r+s)}{\Gamma(r)\Gamma(s)} \int_0^\infty d\lambda \frac{\lambda^{s-1}}{(a + 2b\lambda)^{r+s}}, \tag{3.6}$$

so that Eq. (3.5) can be rewritten as

$$-\left(\frac{8}{3}\right) g^2 \mu^\epsilon \int_0^\infty d\lambda \int \frac{d^n q}{(2\pi)^n} \frac{1}{[q^2 - m^2 + 2\lambda v \cdot (q+p)]^2}. \tag{3.7}$$

Shifting the loop integration momentum by $q \to q - \lambda v$ gives

$$-\left(\frac{8}{3}\right) g^2 \mu^\epsilon \int_0^\infty d\lambda \int \frac{d^n q}{(2\pi)^n} \frac{1}{(q^2 - m^2 - \lambda^2 + 2\lambda v \cdot p)^2}. \tag{3.8}$$

Evaluating Eq. (3.8) using the standard dimensional regularization formula in Eq. (1.44) gives

$$-\left(\frac{8}{3}\right) g^2 \mu^\epsilon \int_0^\infty d\lambda \frac{i}{(4\pi)^{2-\epsilon/2}} \Gamma(\epsilon/2)[\lambda^2 - 2\lambda v \cdot p + m^2]^{-\epsilon/2}. \tag{3.9}$$

The λ integral can be evaluated by using the recursion relation,

$$\begin{aligned} I(a,b,c) &\equiv \int_0^\infty d\lambda (\lambda^2 + 2b\lambda + c)^a \\ &= \frac{1}{1+2a}\left[(\lambda^2 + 2b\lambda + c)^a(\lambda+b)\Big|_0^\infty + 2a(c-b^2) I(a-1,b,c)\right], \end{aligned} \tag{3.10}$$

to convert it to one that is convergent when $\epsilon = 0$,

$$\int_0^\infty d\lambda[\lambda^2 - 2\lambda v\cdot p + m^2]^{-\epsilon/2}$$
$$= \frac{1}{1-\epsilon}\Bigg\{ (\lambda^2 - 2\lambda v\cdot p + m^2)^{-\epsilon/2}(\lambda - v\cdot p)\Big|_0^\infty$$
$$-\epsilon\left[m^2 - (v\cdot p)^2\right]\int_0^\infty d\lambda(\lambda^2 - 2\lambda v\cdot p + m^2)^{-1-\epsilon/2}\Bigg\}. \quad (3.11)$$

The Γ functions in a one-loop dimensionally regularized integral can have at most a $1/\epsilon$ singularity. Since the last term in Eq. (3.11) is multiplied by ϵ, one can set $\epsilon = 0$ in the integrand. The other terms can be evaluated by noting that in dimensional regularization,

$$\lim_{\lambda\to\infty} \lambda^z = 0, \quad (3.12)$$

as long as z depends on ϵ in a way that allows one to analytically continue z to negative values. This gives for Eq. (3.9)

$$-i\frac{g^2}{6\pi^2}(4\pi\mu^2)^{\epsilon/2}\Gamma(\epsilon/2)\frac{1}{1-\epsilon}\Bigg\{(m^2)^{-\epsilon/2}(v\cdot p)$$
$$-\epsilon\left[m^2 - (v\cdot p)^2\right]\int_0^\infty d\lambda(\lambda^2 - 2\lambda v\cdot p + m^2)^{-1}\Bigg\}$$
$$= -i\frac{g^2}{3\pi^2\epsilon}v\cdot p + \text{finite}. \quad (3.13)$$

There is also a tree-level contribution from the counterterm:

$$iv\cdot p(Z_h - 1). \quad (3.14)$$

The sum of Eqs. (3.14) and (3.13) must be finite as $\epsilon \to 0$, so in the $\overline{\text{MS}}$ scheme

$$Z_h = 1 + \frac{g^2}{3\pi^2\epsilon}. \quad (3.15)$$

Note that Z_h is different from the wave-function renormalization of light quark fields given in Eq. (1.86). The anomalous dimension of a heavy quark field is

$$\gamma_h = \frac{1}{2}\frac{\mu}{Z_h}\frac{dZ_h}{d\mu} = -\frac{g^2}{6\pi^2}. \quad (3.16)$$

Composite operators require additional subtractions beyond wave-function renormalization. Consider the heavy-light bare operator

$$O_\Gamma^{(0)} = \bar{q}^{(0)}\Gamma Q_v^{(0)} = \sqrt{Z_q Z_h}\,\bar{q}\Gamma Q_v, \quad (3.17)$$

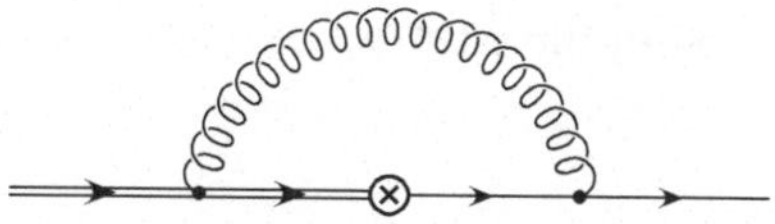

Fig. 3.3. One-loop renormalization of the heavy-light operator $\bar{q}\Gamma Q_v$. The heavy quark is denoted by a double line, the light quark by a single line, and the operator insertion by $\otimes$.

where Γ is any Dirac matrix. The renormalized operator is defined by

$$\begin{aligned} O_\Gamma &= \frac{1}{Z_O} O_\Gamma^{(0)} = \frac{\sqrt{Z_q Z_h}}{Z_O}\, \bar{q}\Gamma Q_v \\ &= \bar{q}\Gamma Q_v + \text{counterterms}, \end{aligned} \tag{3.18}$$

where the additional operator renormalization Z_O can be determined by computing a Green's function with an insertion of O_Γ. For example, Z_O can be determined by considering the one-particle irreducible Green's function of q, $\bar{Q}_v$, and O_Γ. The counterterm in Eq. (3.18) contributes

$$\left(\frac{\sqrt{Z_q Z_h}}{Z_O} - 1 \right) \Gamma \tag{3.19}$$

to this time-ordered product. The one-loop diagram in Fig. 3.3 also gives a divergent contribution to the time-ordered product. Neglecting external momenta (the operator O_Γ contains no derivatives) and using the Feynman gauge, the diagram gives

$$\begin{aligned} &\int \frac{d^n q}{(2\pi)^n} (-ig\mu^{\epsilon/2} T^A) \gamma^\lambda \frac{i \not{q}}{q^2} \Gamma \frac{i}{v \cdot q} (-ig\mu^{\epsilon/2} T^A) v_\lambda \frac{(-i)}{q^2} \\ &\quad = -i \frac{4}{3} g^2 \mu^\epsilon \int \frac{d^n q}{(2\pi)^n} \frac{\not{v} \not{q} \Gamma}{q^4 v \cdot q}. \end{aligned} \tag{3.20}$$

Combining denominators using Eq. (3.6), introducing a gluon mass m to regulate the infrared divergence, and making the change of variables $q \to q - \lambda v$ gives

$$-i \frac{16}{3} g^2 \mu^\epsilon \int d\lambda \int \frac{d^n q}{(2\pi)^n} \frac{\not{v}(\not{q} - \lambda \not{v})\Gamma}{(q^2 - \lambda^2 - m^2)^3}. \tag{3.21}$$

The term proportional to $\not{q}$ is odd in q, and it vanishes on integration. The identity $\not{v}\not{v} = 1$ reduces the remaining integral to be the same as $i/2$ times the derivative of Eq. (3.8) with respect to $v \cdot p$ at $v \cdot p = 0$. Consequently, Fig. 3.3 yields

$$\frac{g^2 \Gamma}{6\pi^2 \epsilon}, \tag{3.22}$$

up to terms that are not divergent as $\epsilon \to 0$. The sum of Eqs. (3.19) and (3.22) must be finite as $\epsilon \to 0$. Using the expressions for $\sqrt{Z_h}$ and $\sqrt{Z_q}$ in Eqs. (3.15) and (1.86) gives

$$Z_O = 1 + \frac{g^2}{4\pi^2\epsilon}, \tag{3.23}$$

and the anomalous dimension is

$$\gamma_O = -\frac{g^2}{4\pi^2}. \tag{3.24}$$

Note that the renormalization of O_Γ is independent of the gamma matrix Γ in the operator. This is a consequence of heavy quark spin symmetry and light quark chiral symmetry, and it is very different from what occurs in the full theory of QCD. For example, in the full theory the operator $\bar{q}_i q_j$ requires renormalization whereas the operator $\bar{q}_i \gamma_\mu q_j$ does not.

As a final example of operator renormalization, consider a composite operator with two heavy quark fields with velocity v and v',

$$T_\Gamma^{(0)} = \bar{Q}_{v'}^{(0)} \Gamma Q_v^{(0)} = Z_h\, \bar{Q}_{v'} \Gamma Q_v. \tag{3.25}$$

The renormalized operator is related to the bare one by means of

$$\begin{aligned} T_\Gamma &= \frac{1}{Z_T} T_\Gamma^{(0)} \\ &= \frac{Z_h}{Z_T} \bar{Q}_{v'} \Gamma Q_v = \bar{Q}_{v'} \Gamma Q_v + \text{counterterms}. \end{aligned} \tag{3.26}$$

One can always choose a frame where $v = v_r$ or where $v' = v_r$, but it is not possible, in general, to go to a frame where both heavy quarks are at rest. Hence T_Γ depends on $w = v \cdot v'$ and we anticipate that its renormalization will also depend on this variable. Heavy quark spin symmetry implies that the renormalization of T_Γ will be independent of Γ. The operator renormalization factor Z_T can be determined from the time-ordered product of $Q_{v'}$, $\bar{Q}_v$ and T_Γ. The counterterm gives the contribution

$$\left(\frac{Z_h}{Z_T} - 1\right)\Gamma, \tag{3.27}$$

and the one-loop Feynman diagram in Fig. 3.4 gives (neglecting external

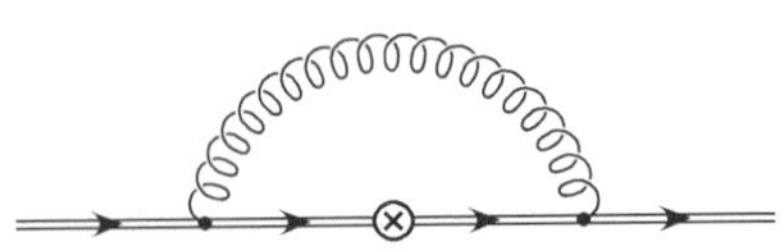

Fig. 3.4. One-loop renormalization of the heavy-heavy operator $\bar{Q}_{v'} \Gamma Q_v$. The heavy quark is denoted by a double line and the operator insertion by $\otimes$.

momenta) the contribution

$$\int \frac{d^n q}{(2\pi)^n}(-igT^A\mu^{\epsilon/2})v'_\lambda(-igT^A\mu^{\epsilon/2})v^\lambda \frac{i}{v'\cdot q}\Gamma\frac{i}{v\cdot q}\frac{(-i)}{q^2}$$
$$= -ig^2\mu^\epsilon\left(\frac{4}{3}\right)w\int\frac{d^n q}{(2\pi)^n}\frac{\Gamma}{q^2(q\cdot v)(q\cdot v')} \tag{3.28}$$

to this three-point function. Using the Feynman trick to first combine the $q \cdot v$ and $q \cdot v'$ terms, and then using Eq. (3.6), gives

$$-ig^2\left(\frac{32}{3}\right)\mu^\epsilon\Gamma w\int_0^\infty d\lambda\int_0^1 dx$$
$$\times\int\frac{d^n q}{(2\pi)^n}\frac{\lambda}{\{q^2+2\lambda[xv+(1-x)v']\cdot q-m^2\}^3}, \tag{3.29}$$

where m has been introduced to regulate the infrared divergence. Performing the q integration by completing the square in the denominator, shifting the q integration and dropping finite terms gives

$$-\frac{g^2}{3\pi^2}\mu^\epsilon w\Gamma\int_0^\infty d\lambda\int_0^1 dx\frac{\lambda}{\{\lambda^2[1+2x(1-x)(w-1)]+m^2\}^{1+\epsilon/2}}, \tag{3.30}$$

where $w = v \cdot v'$. The λ integral can be evaluated explicitly to give

$$-\frac{16}{3}\frac{g^2}{16\pi^2\epsilon}w\Gamma(m^2)^{-\epsilon/2}\int_0^1 dx\frac{1}{[1+2x\,(1-x)\,(w-1)]}. \tag{3.31}$$

Performing the x integral yields for the part proportional to $1/\epsilon$,

$$-\left(\frac{16}{3}\right)\frac{g^2}{16\pi^2\epsilon}w\,r(w)\,\Gamma, \tag{3.32}$$

where

$$r(w) = \frac{1}{\sqrt{w^2-1}}\ln\left(w+\sqrt{w^2-1}\right), \tag{3.33}$$

Demanding that the sum of Eq. (3.27) and Eq. (3.32) be finite as $\epsilon \to 0$ determines the operator renormalization factor Z_T. Using Eq. (3.15) we find that

$$Z_T = 1 - \frac{g^2}{3\pi^2\epsilon}[w\,r(w)-1], \tag{3.34}$$

and the operator anomalous dimension is

$$\gamma_T = \frac{g^2}{3\pi^2}[w\,r(w)-1]. \tag{3.35}$$

Note that the renormalization of $T_\Gamma = \bar{Q}_{v'}\Gamma Q_v$ depends on the dot product of four velocities $w = v \cdot v'$. This is reasonable since Q_v is a different field for each value of the four-velocity. At the zero-recoil point $w = 1$ the operator $\bar{Q}_v \gamma_\mu Q_v$ is a conserved current associated with heavy quark flavor symmetry and hence is not renormalized. The anomalous dimension γ_T near $w = 1$ has the expansion

$$\gamma_T = \frac{g^2}{\pi^2} \left[\frac{2}{9} (w-1) - \frac{1}{15} (w-1)^2 + \cdots \right], \tag{3.36}$$

and vanishes at $w = 1$.

3.2 Matching between QCD and HQET

The computation of physical quantities in QCD using HQET requires relating QCD operators to HQET operators, which is referred to as "matching." Consider the QCD vector current operator,

$$V_\nu = \bar{q} \gamma_\nu Q, \tag{3.37}$$

involving a heavy quark field Q and a light quark field q. Matrix elements of this operator are important for semileptonic decays such as $\bar{B} \to \pi e \bar{\nu}_e$ and $D \to \pi \bar{e} \nu_e$. In QCD this operator is not renormalized, since it is conserved in the limit that the (heavy and light) quark masses vanish. Quark mass terms are dimension-three operators, and therefore do not affect anomalous dimensions. Matrix elements of the full QCD vector current between physical states contain large logarithms of the quark mass m_Q divided by a typical hadronic momentum, which is of the order of Λ_{QCD}. These logarithms can be resummed using HQET. In HQET, matrix elements of operators renormalized at μ can only contain logarithms of Λ_{QCD}/μ. There are no logarithms of m_Q/μ, since HQET makes no reference to the large-momentum scale m_Q. The logarithms of m_Q/μ are obtained by scaling the HQET operators between m_Q and μ, using the anomalous dimensions computed in the previous section.

The first step in computing matrix elements of V_ν is to relate the QCD operator to HQET operators. One can do this by computing matrix elements of the QCD operator between quarks at a scale μ, and comparing this with matrix elements of HQET operators renormalized at the same scale. Both calculations are done in perturbation theory, and are in general infrared divergent. However, the matching conditions depend on the difference between the computations in QCD and HQET. Since HQET is constructed to reproduce the low-momentum dynamics of QCD, the infrared divergences cancel in the matching conditions. One can therefore compute the matching conditions by using any convenient infrared regulator. It is crucial that the matching conditions do not depend on infrared effects; otherwise they would depend on the nonperturbative scale Λ_{QCD}, and they would not be computable by using perturbation theory. Two common ways to regulate infrared divergences are to use a gluon mass and to use dimensional regularization. In this chapter, we will use dimensional regularization. If the scale

μ is chosen to be of the order of the heavy quark mass m_Q, the computation of the matching between the full and effective theories will be an expansion in $\alpha_s(\mu)$, with no large logarithms. For the specific example of the heavy $\to$ light vector current, this expansion takes the form

$$V^\lambda = C_1^{(V)}\left[\frac{m_Q}{\mu}, \alpha_s(\mu)\right]\bar{q}\gamma^\lambda Q_v + C_2^{(V)}\left[\frac{m_Q}{\mu}, \alpha_s(\mu)\right]\bar{q}v^\lambda Q_v. \tag{3.38}$$

The right-hand side of Eq. (3.38) includes all dimension-three operators with the same quantum numbers as the vector current V^λ. Higher dimension operators are suppressed by powers of $1/m_Q$. They can also be computed in a systematic expansion to determine the $1/m_Q$ corrections, as will be discussed in Chapter 4. Other dimension-three operators can be rewritten in terms of the two operators given above. For example, $\bar{q}i\sigma^{\mu\nu}v_\nu Q_v = -(1/2)\bar{q}(\gamma^\mu \not{v} - \not{v}\gamma^\mu)Q_v = -\bar{q}\gamma^\mu Q_v + v^\mu \bar{q} Q_v$, and so is not a linearly independent operator.

The matching calculation between QCD and HQET at the scale m_Q determines $C_i^{(V)}[1, \alpha_s(m_Q)]$. At lowest order in α_s (tree level), the matching condition is trivial,

$$\begin{aligned} C_1^{(V)}[1, \alpha_s(m_Q)] &= 1 + \mathcal{O}[\alpha_s(m_Q)], \\ C_2^{(V)}[1, \alpha_s(m_Q)] &= \mathcal{O}[\alpha_s(m_Q)], \end{aligned} \tag{3.39}$$

since at tree level, the field Q can be replaced by Q_v up to corrections of the order of $1/m_Q$. The one-loop corrections to $C_i^{(V)}$ will be computed in Sec. 3.3.

In the general case, one has a QCD operator $O_{\rm QCD}$ renormalized at the scale m_Q, which can be expressed as a linear combination of HQET operators O_i renormalized at the scale μ,

$$O_{\rm QCD}(m_Q) = \sum_i C_i\left[\frac{m_Q}{\mu}, \alpha_s(\mu)\right] O_i(\mu), \tag{3.40}$$

where the coefficients $C_i[1, \alpha_s(\mu)]$ are computed by doing a perturbative matching condition calculation at the scale $\mu = m_Q$. One can then obtain the coefficients $C_i[m_Q/\mu, \alpha_s(\mu)]$ at some lower scale $\mu < m_Q$ by renormalization group scaling in the effective theory, using the same procedure as that used for the weak Hamiltonian in Sec. 1.6. The operators O_i satisfy the renormalization group equation in Eq. (1.129). Since the left-hand side of Eq. (3.40) is μ independent, this implies that the coefficients satisfy the renormalization group equation shown in Eq. (1.133), with the solution given by Eq. (1.134).

The renormalization group equation solution in Eq. (1.134) can be written out explicitly in the case in which a single operator is multiplicatively renormalized, so that γ is a number rather than a matrix. The anomalous dimension, β function, and matching coefficient have the perturbative expansions

$$\begin{aligned}
\gamma(g) &= \gamma_0 \frac{g^2}{4\pi} + \gamma_1 \left(\frac{g^2}{4\pi}\right)^2 + \cdots \\
\beta(g) &= -\beta_0 \frac{g^3}{4\pi} - \beta_1 \frac{g^5}{(4\pi)^2} + \cdots, \\
C[1, \alpha_s(m_Q)] &= C_0 + C_1 \alpha_s(m_Q) + \cdots .
\end{aligned} \tag{3.41}$$

Integrating Eq. (1.134) gives

$$\begin{aligned}
C\left[\frac{m_Q}{\mu}, \alpha_s(m_Q)\right] &= [C_0 + C_1\alpha_s(m_Q) + \cdots] \\
&\quad \times \left\{\exp \int_{g(\mu)}^{g(m_Q)} \frac{dg}{g} \left[\frac{\gamma_0}{\beta_0} + \left(\frac{\gamma_1}{\beta_0} - \frac{\gamma_0\beta_1}{\beta_0^2}\right)\frac{g^2}{4\pi} + \cdots\right]\right\} \\
&= \left[\frac{\alpha_s(\mu)}{\alpha_s(m_Q)}\right]^{-(\gamma_0/2\beta_0)} \\
&\quad \times \left\{C_0 + C_0\left(\frac{\gamma_1}{2\beta_0} - \frac{\gamma_0\beta_1}{2\beta_0^2}\right)[\alpha_s(m_Q) - \alpha_s(\mu)] + C_1\alpha_s(m_Q) + \cdots\right\}.
\end{aligned} \tag{3.42}$$

The terms explicitly displayed in this equation sum all subleading logarithms of the form $\alpha_s^{n+1} \ln^n(m_Q/\mu)$. To evaluate the subleading logarithms requires knowing the two-loop anomalous dimension and β function, and the one-loop matching coefficient C_1. The two-loop β function is scheme independent, but C_1 and γ_1 are both scheme dependent in general. Retaining only the one-loop anomalous dimension γ_0 and the one-loop β function β_0 sums all the leading logarithms $\alpha_s^n \ln^n(m_Q/\mu)$.

The leading logarithms can be summed in the case of operator mixing by diagonalizing the anomalous dimension matrix γ_0, and then using Eq. (3.42). The two-loop equations with operator mixing cannot be simplified in the same way, because in general, γ_0 and γ_1 cannot be simultaneously diagonalized, and the equation has to be integrated numerically.

It should now be clear how to interpret the predictions for heavy meson decay constants and form factors obtained in Secs. 2.8–2.11. For the decay constants, the coefficient a is subtraction-point dependent, and Eq. (2.62) holds up to perturbative matching corrections when a is evaluated at $\mu = m_Q$. The μ dependence of a is determined by the anomalous dimension in Eq. (3.24). The situation is similar for the Isgur-Wise functions that occur in $\bar{B} \to D^{(*)} e\bar{\nu}_e$ and $\Lambda_b \to \Lambda_c e\bar{\nu}_e$ decays. The Isgur-Wise functions are matrix elements of HQET operators and also depend on the subtraction point μ due to the anomalous dimension in Eq. (3.35). The expression for the form factors in terms of the Isgur-Wise functions are valid up to perturbative matching corrections provided the Isgur-Wise functions are evaluated at a subtraction point around $m_{c,b}$, e.g.,

$\mu = \sqrt{m_c m_b}$. Note, however, that the anomalous dimension γ_T vanishes at $w=1$, and therefore the normalization conditions $\xi(1)=1$ and $\zeta(1)=1$ in Eqs. (2.93) and (2.109) are μ independent.

3.3 Heavy-light currents

The tree-level matching conditions for heavy $\to$ light currents are given in Eq. (3.39). The one-loop corrections to this result can be determined by computing at order α_s a matrix element of the left-hand side of Eq. (3.38) in the full theory of QCD and equating it with the corresponding matrix element of the right-hand side of Eq. (3.38) calculated in HQET. A convenient matrix element is that between an on-shell heavy quark with four-momentum $p = m_Q v$ as the initial state and an on-shell massless quark state with four-momentum zero as the final state. These are not physical states since the strong interactions confine. However, Eq. (3.38) holds at the operator level and so these unphysical states can be used to determine the matching coefficients, $C_1^{(V)}$ and $C_2^{(V)}$.

The order α_s matrix element in QCD contains the one-loop vertex correction, as well as the one-loop correction to the propagator for the heavy and light quark fields. The quark propagators have the form [analytic $+\, iR^{(Q)}/(\not{p} - m_Q)$] and [analytic $+\, iR^{(q)}/\not{p}$] near the poles $p^2 = m_Q^2$ and $p^2 = 0$, respectively. The residues $R^{(Q)}$ and $R^{(q)}$ have perturbative expansions

$$R^{(Q)} = 1 + R_1^{(Q)} \alpha_s(\mu) + \cdots \tag{3.43}$$

and

$$R^{(q)} = 1 + R_1^{(q)} \alpha_s(\mu) + \cdots . \tag{3.44}$$

The desired matrix element in full QCD is obtained from the LSZ reduction formula,

$$\langle q(0, s')| V^\lambda | Q(p, s)\rangle = \left[R^{(Q)} R^{(q)}\right]^{1/2} \bar{u}(0, s')\left[\gamma^\lambda + V_1^\lambda \alpha_s(\mu)\right] u(p, s), \tag{3.45}$$

where γ^λ is the tree-level vertex, and $\alpha_s V_1^\lambda$ is the one-loop correction to the vertex from Fig. 1.4. The one-loop correction to the vertex has the expansion $(p = m_Q v)$

$$V_1^\lambda = V_1^{(1)} \gamma^\lambda + V_1^{(2)} v^\lambda, \tag{3.46}$$

as will be shown in Eq. (3.65).

The expression for the analogous matrix element in HQET is

$$\langle q(0, s')| \bar{q} \Gamma Q_v | Q(v, s)\rangle = \left[R^{(h)} R^{(q)}\right]^{1/2} \bar{u}(0, s')\left[1 + V_1^{\text{eff}} \alpha_s(\mu)\right] \Gamma u(0, s), \tag{3.47}$$

where $R^{(h)}$ is the the residue of the heavy quark propagator near its pole, $iR^{(h)}/p \cdot v +$ analytic, and $\alpha_s V_1^{\text{eff}} \Gamma$ is the one-loop vertex correction in Fig. 3.3, which is independent of the Γ matrix structure of the operator $\bar{q}\Gamma Q_v$.

Comparing Eqs. (3.45)–(3.47) and (3.38) gives

$$C_1^{(V)}\left[\frac{m_Q}{\mu}, \alpha_s(\mu)\right] = 1 + \left\{\frac{1}{2}\left[R_1^{(Q)} - R_1^{(h)}\right] + V_1^{(1)} - V_1^{\text{eff}}\right\} \alpha_s(\mu) + \cdots,$$
$$C_2^{(V)}\left[\frac{m_Q}{\mu}, \alpha_s(\mu)\right] = V_1^{(2)} \alpha_s(\mu) + \cdots, \tag{3.48}$$

where the ellipses denote terms higher order in $\alpha_s(\mu)$. $R_1^{(q)}$ does not occur in Eqs. (3.48) because it is common to both the HQET and full QCD calculations of the matrix element. The quantities R_1 and V_1 are ultraviolet finite as $\epsilon \to 0$ but they have infrared divergences, which must be regulated before computing these quantities. The coefficients $C_1^{(V)}$ and $C_2^{(V)}$ are not infrared divergent, so the infrared divergence cancels in the matching condition, which involves differences $R_1^{(Q)} - R_1^{(h)}$ and $V_1^{(1)} - V_1^{\text{eff}}$ in the full and effective theories. It is important to use the same infrared regulator in both theories when computing matching conditions.

In this section, dimensional regularization will be used to regulate both the infrared and ultraviolet divergences. All graphs are computed in $4 - \epsilon$ dimensions, and the limit $\epsilon \to 0$ is taken at the end of the computation. Graphs will have $1/\epsilon$ poles, which arise from ultraviolet and infrared divergences. Only the $1/\epsilon$ ultraviolet divergences are canceled by counterterms. As a simple example, consider the integral

$$\int \frac{d^n q}{(2\pi)^n} \frac{1}{q^4} = 0. \tag{3.49}$$

The integral is ultraviolet and infrared divergent, but it is zero when evaluated in dimensional regularization. The infrared divergence can be regulated by introducing a mass to give

$$\int \frac{d^n q}{(2\pi)^n} \frac{1}{(q^2 - m^2)^2} = \frac{i}{8\pi^2 \epsilon} + \text{finite}. \tag{3.50}$$

Thus the original integral can be written as

$$\int \frac{d^n q}{(2\pi)^n} \frac{1}{q^4} = \frac{i}{8\pi^2 \epsilon} - \frac{i}{8\pi^2 \epsilon}, \tag{3.51}$$

where the first term is the ultraviolet divergence, and the second term is the infrared divergence. The counterterm contribution to the integral is $-i/8\pi^2\epsilon$, which cancels the ultraviolet divergence and leaves

$$\int \frac{d^n q}{(2\pi)^n} \frac{1}{q^4} + \text{counterterm} = -\frac{i}{8\pi^2 \epsilon}, \tag{3.52}$$

where the right-hand side now only has an infrared divergence.

3.3.1 The QCD computation

The two-point function of renormalized heavy quark fields in the full QCD theory gets two contributions at order α_s. One is the one-loop diagram in Fig. 1.2 denoted by the subscript fd, and the other is the tree-level matrix element of the counterterm that cancels the $1/\epsilon$ ultraviolet divergence, denoted by the subscript ct. In the Feynman gauge, the one-loop contribution in Fig. 1.2 gives the quark self-energy Σ_{fd},

$$\begin{aligned} -i\Sigma_{\text{fd}} &= \int \frac{d^n q}{(2\pi)^n}\left(-igT^A\mu^{\epsilon/2}\right)\gamma^\alpha \frac{i(\not p + \not q + m_Q)}{\left[(p+q)^2 - m_Q^2\right]}\left(-igT^A\mu^{\epsilon/2}\right)\gamma_\alpha \frac{(-i)}{q^2} \\ &= -g^2\left(\frac{4}{3}\right)\mu^\epsilon \int \frac{d^n q}{(2\pi)^n}\frac{\gamma^\alpha(\not q + \not p)\gamma_\alpha + n m_Q}{q^2\left[(q+p)^2 - m_Q^2\right]}. \end{aligned} \tag{3.53}$$

Using the identity $\gamma^\alpha\gamma_\mu\gamma_\alpha = 2\gamma_\mu - \gamma^\alpha\gamma_\alpha\gamma_\mu = (2-n)\gamma_\mu$ and combining denominators gives

$$\begin{aligned} -i\Sigma_{\text{fd}} &= -g^2\left(\frac{4}{3}\right)\mu^\epsilon \int_0^1 dx \int \frac{d^n q}{(2\pi)^n}\frac{(2-n)(\not q + \not p) + n m_Q}{\left[q^2 + 2q\cdot p x - m_Q^2 x + p^2 x\right]^2} \\ &= -g^2\left(\frac{4}{3}\right)\mu^\epsilon \int_0^1 dx \int \frac{d^n q}{(2\pi)^n}\frac{(2-n)(1-x)\not p + n m_Q}{\left[q^2 + p^2 x(1-x) - m_Q^2 x\right]^2}. \end{aligned} \tag{3.54}$$

The self-energy has the form

$$\Sigma(p) = A(p^2)m_Q + B(p^2)\not p. \tag{3.55}$$

Since the full propagator is $i/[\not p - m_Q - \Sigma(p)]$, it is straightforward to see that the residue at the pole is

$$R_1^{(Q)}\alpha_s(\mu) = B\left(m_Q^2\right) + 2m_Q^2\left.\frac{\mathrm{d}(A+B)}{\mathrm{d}p^2}\right|_{p^2=m_Q^2}. \tag{3.56}$$

Performing the $d^n q$ integration in Eq. (3.54) yields the following expressions for A and B:

$$\begin{aligned} A_{\text{fd}}(p^2) &= \frac{g^2}{12\pi^2}(4\pi\mu^2)^{\epsilon/2}\Gamma(\epsilon/2)(4-\epsilon)\int_0^1 dx\left[m_Q^2 x - p^2 x(1-x)\right]^{-\epsilon/2}, \\ B_{\text{fd}}(p^2) &= -\frac{g^2}{12\pi^2}(4\pi\mu^2)^{\epsilon/2}\Gamma(\epsilon/2)(2-\epsilon) \\ &\quad\times\int_0^1 dx(1-x)\left[m_Q^2 x - p^2 x(1-x)\right]^{-\epsilon/2}. \end{aligned} \tag{3.57}$$

The on-shell renormalization factor R_1 of Eq. (3.56) can be obtained by substituting for A and B and integrating over x, using the identity

$$\int_0^1 x^a(1-x)^b = \frac{\Gamma(1+a)\Gamma(1+b)}{\Gamma(2+a+b)}. \tag{3.58}$$

Expanding around $\epsilon=0$ gives

$$R_{1,\mathrm{fd}}\alpha_s = -\frac{g^2}{12\pi^2}\left(\frac{6}{\epsilon}+4-3\gamma+3\ln\frac{4\pi\mu^2}{m_Q^2}\right). \tag{3.59}$$

The $1/\epsilon$ terms include both the infrared and ultraviolet divergences. The counterterm contribution is $-i\Sigma_{\mathrm{ct}}=i(Z_q-1)\not p - i(Z_m-1)m$, i.e., $A_{\mathrm{ct}}=(Z_m-1)$ and $B_{\mathrm{ct}}=-(Z_q-1)$, which gives the counterterm contribution to $R_{1,\mathrm{ct}}\alpha_s$ of $-(Z_q-1)$. Adding this [from Eq. (1.86)] to $R_{1,\mathrm{fd}}\alpha_s$ and rescaling $4\pi\mu^2\to\mu^2e^{\gamma}$ to convert to the $\overline{\mathrm{MS}}$ scheme gives the final result,

$$R_1^{(Q)}\alpha_s = -\frac{g^2}{12\pi^2}\left(\frac{4}{\epsilon}+4+3\ln\frac{\mu^2}{m_Q^2}\right), \tag{3.60}$$

where the $1/\epsilon$ divergence in Eq. (3.60) is only an infrared divergence.

Next, consider the order α_s contribution to the one-particle irreducible vertex in full QCD shown in Fig. 1.4. In the Feynman gauge the graph gives

$$\int\frac{d^nq}{(2\pi)^n}(-ig\mu^{\epsilon/2}T^A)\gamma_\alpha\frac{i\not q}{q^2}\gamma^\lambda i\frac{(\not p+\not q+m_Q)}{\left[(p+q)^2-m_Q^2\right]}(-ig\mu^{\epsilon/2}T^A)\gamma^\alpha\frac{(-i)}{q^2}. \tag{3.61}$$

Combining denominators, shifting the integration variable $q\to q-px$, and using $p^2=m_Q^2$ gives

$$-ig^2\mu^\epsilon\left(\frac{8}{3}\right)\int_0^1 dx(1-x)\int\frac{d^nq}{(2\pi)^n}\frac{1}{\left(q^2-m_Q^2x^2\right)^3}$$
$$\times\{\gamma_\alpha(\not q-\not px)\gamma^\lambda[\not q+\not p(1-x)]\gamma^\alpha+m_Q\gamma_\alpha(\not q-\not px)\gamma^\lambda\gamma^\alpha\}. \tag{3.62}$$

The numerator can be simplified using the relations $\gamma_\alpha\not a\not b\not c\gamma^\alpha = -2\not c\not b\not a-(n-4)\not a\not b\not c$, and $\gamma_\alpha\not a\not b\gamma^\alpha=4a\cdot b+(n-4)\not a\not b$. Terms odd in q vanish on integration. Terms involving $\not p$ can be simplified by anticommuting $\not p$ through any γ matrices until it is at the right, where it can be eliminated using $\not p=m_Q$ when acting on the heavy quark spinor. The final expression is

$$-ig^2\mu^\epsilon\left(\frac{8}{3}\right)\int_0^1 dx(1-x)\int\frac{d^nq}{(2\pi)^n}\frac{1}{\left(q^2-m_Q^2x^2\right)^3}$$
$$\times\left\{\frac{q^2}{n}(2-n)^2\gamma^\lambda-2m_Qp^\lambda(n-2)x^2+m_Q^2\gamma^\lambda x[x(n-2)-2]\right\}. \tag{3.63}$$

Evaluating the q integrals and using $p = m_Q v$ gives

$$\frac{g^2}{12\pi^2}(4\pi\mu^2)^{\epsilon/2}\int_0^1 dx(1-x)\left(m_Q^2 x^2\right)^{-\epsilon/2}\left\{\frac{1}{2}\Gamma(\epsilon/2)(2-\epsilon)^2\gamma^\lambda\right.$$
$$\left.+2\Gamma(1+\epsilon/2)v^\lambda(2-\epsilon)-\Gamma(1+\epsilon/2)\gamma^\lambda\frac{1}{x}\left[x(2-\epsilon)-2\right]\right\}. \quad (3.64)$$

Evaluating the x integral and expanding in ϵ gives

$$\frac{g^2}{12\pi^2}\{-2\gamma^\lambda + 2v^\lambda\}. \quad (3.65)$$

The counterterm contribution is determined by the renormalization of the current $\bar{q}\gamma^\lambda Q$ in QCD. Since this is a partially conserved current (i.e., is conserved in the limit that the masses vanish), it is not renormalized. The only remaining counterterm contribution is the QCD wave-function renormalization $Z_q - 1 = -2\alpha_s/3\pi\epsilon$ to $V_1^{(1)}\alpha_s$, from Eq. (1.86). Adding this to Eq. (3.65) gives

$$V_1^{(1)}\alpha_s = -\frac{2\alpha_s}{3\pi}\left(\frac{1}{\epsilon}+1\right),$$
$$V_1^{(2)}\alpha_s = \frac{2\alpha_s}{3\pi}. \quad (3.66)$$

3.3.2 The HQET computation

We have now calculated all the quantities in full QCD that occur in Eq. (3.48) for $C_1^{(V)}$ and $C_2^{(V)}$. It remains to calculate the HQET quantities. In the Feynman gauge the HQET heavy quark self-energy obtained from the Feynman diagram in Fig. 3.2 is

$$-i\Sigma_{\rm fd}(p) = -\left(\frac{4}{3}\right)g^2\mu^\varepsilon\int\frac{d^n q}{(2\pi)^n}\frac{1}{q^2\, v\cdot(p+q)}, \quad (3.67)$$

The residue at the pole is

$$R_1^{(h)}\alpha_s = v^\alpha\left.\frac{\partial\Sigma}{\partial p^\alpha}\right|_{p\cdot v=0}. \quad (3.68)$$

Evaluating Eq. (3.67) by combining denominators, the q integral gives

$$-i\Sigma_{\rm fd} = -i\frac{g^2}{6\pi^2}(4\pi\mu^2)^{\epsilon/2}\Gamma(\epsilon/2)\int_0^\infty d\lambda(\lambda^2-2\lambda p\cdot v)^{-\epsilon/2}$$
$$= -i\frac{g^2}{6\pi^2}(4\pi\mu^2)^{\epsilon/2}(-p\cdot v)^{1-\epsilon}\frac{\Gamma(\epsilon/2)\Gamma(1-\epsilon/2)\Gamma(-1/2+\epsilon/2)}{2\sqrt{\pi}}. \quad (3.69)$$

This yields $R_{1,\text{fd}}^{(h)} = 0$, since $\lim_{p\to 0}(-p\cdot v)^{-\epsilon} = 0$. The only contribution to $R_1^{(h)}$ is $-(Z_h - 1)$ from the counterterm, Eq. (3.14),

$$R_1^{(h)}\alpha_s = R_{1,\text{ct}}^{(h)}\alpha_s = -\frac{4\alpha_s}{3\pi\epsilon}. \tag{3.70}$$

The vertex calculation is also much simpler in HQET than in full QCD. The Feynman diagram in Fig. 3.3 gives

$$-ig^2\mu^\epsilon\left(\frac{4}{3}\right)\int\frac{d^nq}{(2\pi)^n}\frac{\not{v}\not{q}\Gamma}{(q^2)^2 v\cdot q}. \tag{3.71}$$

Combining denominators and evaluating the q integral gives

$$\frac{g^2}{6\pi^2}\Gamma(4\pi\mu^2)^{\epsilon/2}\Gamma(1+\epsilon/2)\int_0^\infty d\lambda\,\lambda^{-1-\epsilon}, \tag{3.72}$$

which is zero in dimensional regularization. The only contribution is from the counterterm, the negative of Eq. (3.22), which implies that

$$V_1^{\text{eff}}\alpha_s = -\frac{2\alpha_s}{3\pi\epsilon}. \tag{3.73}$$

Putting the pieces Eqs. (3.48), (3.60), (3.66), (3.70), and (3.73) of the matching calculation together yields

$$\begin{aligned} C_1^{(V)}\left[\frac{m_Q}{\mu},\alpha_s(\mu)\right] &= 1 + \frac{\alpha_s(\mu)}{\pi}\left[\ln(m_Q/\mu) - \frac{4}{3}\right], \\ C_2^{(V)}\left[\frac{m_Q}{\mu},\alpha_s(\mu)\right] &= \frac{2}{3}\frac{\alpha_s(\mu)}{\pi}. \end{aligned} \tag{3.74}$$

All the $1/\epsilon$ infrared divergences have canceled in the matching conditions. Note that in $C_1^{(V)}$ there is a logarithm of (m_Q/μ). That is why in our initial condition for the $C^{(V)}$'s we took $\mu = m_Q$. If μ was chosen very different from m_Q, large logarithms would prevent a perturbative evaluation of the initial values for the $C^{(V)}$'s. Of course, we do not have to pick $\mu = m_Q$ precisely. One may just as well use $\mu = m_Q/2$ or $\mu = 2m_Q$, for example. The μ dependence of the coefficients $C_i^{(V)}$ is connected with the anomalous dimension of the HQET operator $\bar{q}\gamma^\lambda Q_v$. Here $\mu[dC_1^{(V)}/d\mu]$ is the anomalous dimension γ_O given in Eq. (3.24). The absence of a logarithm in $C_2^{(V)}$ shows explicitly that $\bar{q}\gamma^\lambda Q_v$ does not mix with $\bar{q}v^\lambda Q_v$, which is consistent with our expectations based on spin and chiral symmetries.

A similar matching condition holds for the axial current, $A^\mu = \bar{q}\gamma^\mu\gamma_5 Q$.

$$A^\mu = C_1^{(A)}\left[\frac{m_Q}{\mu}, \alpha_s(\mu)\right]\bar{q}\gamma^\mu\gamma_5 Q_v + C_2^{(A)}\left[\frac{m_Q}{\mu}, \alpha_s(\mu)\right]\bar{q}v^\mu\gamma_5 Q_v. \quad (3.75)$$

It is simple to deduce the $C_j^{(A)}$, given our calculation of the $C_j^{(V)}$'s. Rewrite the axial current as $A^\mu = -\bar{q}\gamma_5\gamma^\mu Q$. γ_5 acting on the massless quark q gives $\pm$ depending on the chirality of the quark. Chirality is conserved by the gluon vertices, so the calculation of matching conditions proceeds just as in the vector current case, except that $\bar{q}$ should be replaced everywhere by $\bar{q}\gamma_5$. At the end of the calculation, the γ_5 is moved back next to Q_v, producing a compensating minus sign for $\gamma^\mu\gamma_5$, but not for $v^\mu\gamma_5$. Thus

$$C_1^{(A)}\left[\frac{m_Q}{\mu}, \alpha_s(\mu)\right] = C_1^{(V)}\left[\frac{m_Q}{\mu}, \alpha_s(\mu)\right], \quad (3.76)$$

$$C_2^{(A)}\left[\frac{m_Q}{\mu}, \alpha_s(\mu)\right] = -C_2^{(V)}\left[\frac{m_Q}{\mu}, \alpha_s(\mu)\right]. \quad (3.77)$$

The results of this section can be used to compute the α_s corrections to the pseudoscalar and vector meson decay constant relations given in Sec. 2.8. The QCD vector and axial current operators match the linear combination of HQET operators given in Eqs. (3.38) and (3.75). Computing the matrix elements of the HQET operators $\bar{q}\Gamma^\mu Q_v$ (renormalized at μ) as in Eq. (2.63) gives

$$a(\mu) \times \begin{cases} -iv^\mu P_v^{(Q)} & \text{if } \Gamma^\mu = \gamma^\mu\gamma_5, \\ iv^\mu P_v^{(Q)} & \text{if } \Gamma^\mu = v^\mu\gamma_5, \\ P_v^{*(Q)\mu} & \text{if } \Gamma^\mu = \gamma^\mu, \\ 0 & \text{if } \Gamma^\mu = v^\mu. \end{cases} \quad (3.78)$$

Combining this with the matching conditions gives

$$\begin{aligned} f_{P^*} &= \sqrt{m_{P*}}\, a(\mu) C_1^{(V)}(\mu), \\ f_P &= \frac{1}{\sqrt{m_P}}\, a(\mu)\left[C_1^{(A)}(\mu) - C_2^{(A)}(\mu)\right]. \end{aligned} \quad (3.79)$$

The μ dependence of the matrix element $a(\mu)$ is given by the anomalous dimension of the heavy-light operators, Eq. (3.24),

$$\mu\frac{da}{d\mu} = -\gamma_O a = \frac{\alpha_s}{\pi}a. \quad (3.80)$$

This μ dependence is canceled by the μ dependence in the coefficients $C_i^{(V,A)}$, so that the complete answer for the measurable quantity $f_{P,P*}$ is μ independent.

For example,

$$\sqrt{m_P}\,\mu\frac{\mathrm{d}f_P}{\mathrm{d}\mu} = \mu\frac{\mathrm{d}a}{\mathrm{d}\mu}\left[C_1^{(A)} - C_2^{(A)}\right] + a\,\mu\frac{\mathrm{d}}{\mathrm{d}\mu}\left[C_1^{(A)} - C_2^{(A)}\right]$$
$$= \frac{\alpha_s}{\pi}\,a\left[C_1^{(A)} - C_2^{(A)}\right] + a\left(-\frac{\alpha_s}{\pi} + 0\right)$$
$$= 0 + \mathcal{O}(\alpha_s^2). \tag{3.81}$$

Equation (3.79) gives the α_s correction to the ratio of the pseudoscalar and vector meson decay constants,

$$\frac{f_{P^*}}{f_P} = \sqrt{m_{P^*}m_P}\left[\frac{C_1^{(V)}}{C_1^{(A)} - C_2^{(A)}}\right] = \sqrt{m_{P^*}m_P}\left[1 - \frac{2}{3}\frac{\alpha_s(m_Q)}{\pi}\right]. \tag{3.82}$$

The α_s correction to the ratio of pseudoscalar meson decay constants for the D and B mesons can also be determined. Heavy quark flavor symmetry implies that $a(\mu)$, the matrix element in the effective theory, is independent of the quark mass. The matching from QCD to the effective theory is done at the scale $m_Q = m_b$ for the $\bar{B}$ meson system, and $m_Q = m_c$ for the D meson system. This determines

$$\frac{f_B\sqrt{m_B}}{f_D\sqrt{m_D}} = \left[\frac{a(m_b)}{a(m_c)}\right]\frac{C_1^{(A)}[1,\alpha_s(m_b)] - C_2^{(A)}[1,\alpha_s(m_b)]}{C_1^{(A)}[1,\alpha_s(m_c)] - C_2^{(A)}[1,\alpha_s(m_c)]}$$
$$= \left[\frac{\alpha_s(m_b)}{\alpha_s(m_c)}\right]^{-6/25}$$
$$\times\left\{1 + [\alpha_s(m_b) - \alpha_s(m_c)]\left[-\frac{2}{3\pi} + \left(\frac{\gamma_{1O}}{2\beta_0} - \frac{\gamma_{0O}\beta_1}{2\beta_0^2}\right)\right]\right\}. \tag{3.83}$$

To complete the prediction for the ratio of B and D meson decay constants, the two-loop correction to the anomalous dimension of O_Γ, γ_{1O}, and the two-loop contribution to the β function, β_1, are needed. These can be found in the literature. The leading logarithmic prediction for the ratio of B and D meson decay constants is

$$\frac{f_B\sqrt{m_B}}{f_D\sqrt{m_D}} = \left[\frac{\alpha_s(m_b)}{\alpha_s(m_c)}\right]^{-6/25}. \tag{3.84}$$

The matching conditions in this section have been computed keeping the $1/\epsilon$ infrared divergent quantities, to show explicitly that the divergences cancel in the matching coefficients. This cancellation provides a useful check on the calculation. The matching conditions can be computed more simply if one is willing to forego this check. One can simply compute only the finite parts of the dimensionally regulated graphs in the full and effective theory to compute the matching conditions. The $1/\epsilon$ ultraviolet divergences are canceled by counterterms, and the $1/\epsilon$ infrared divergences will cancel in the matching conditions, and so need not be retained. One also need not compute any diagrams in the effective theory, since all on-shell graphs in the effective theory vanish on dimensional regularization. We saw this explicitly in Eqs. (3.69) and (3.72). The reason is that graphs that contain no dimensionful parameter vanish in dimensional regularization.

Since m_b/m_c is not very large, there is no reason to sum the leading logarithms of m_b/m_c. If one matches onto HQET simultaneously for the b and c quarks at a scale μ, then Eqs. (3.74), (3.76), and (3.77) imply that

$$\frac{f_B\sqrt{m_B}}{f_D\sqrt{m_D}} = 1 + \frac{\alpha_s(\mu)}{\pi}\ln\left(\frac{m_b}{m_c}\right). \tag{3.85}$$

Eq. (3.85) can also be derived by expanding Eq. (3.84) to order α_s.

3.4 Heavy-heavy currents

$\bar{B} \to D^{(*)}e\bar{\nu}_e$ and $\Lambda_b \to \Lambda_c e\bar{\nu}_e$ decay rates are determined by matrix elements of the vector current, $\bar{c}\gamma_\mu b$, and the axial vector current $\bar{c}\gamma_\mu\gamma_5 b$. The matching of these currents in full QCD onto operators in HQET has the form

$$\begin{aligned}
\bar{c}\gamma_\mu b = \;& C_1^{(V)}\left[\frac{m_b}{\mu}, \frac{m_c}{\mu}, \alpha_s(\mu), w\right]\bar{c}_{v'}\gamma_\mu b_v \\
&+ C_2^{(V)}\left[\frac{m_b}{\mu}, \frac{m_c}{\mu}, \alpha_s(\mu), w\right]\bar{c}_{v'}v_\mu b_v \\
&+ C_3^{(V)}\left[\frac{m_b}{\mu}, \frac{m_c}{\mu}, \alpha_s(\mu), w\right]\bar{c}_{v'}v'_\mu b_v
\end{aligned} \tag{3.86}$$

and

$$\begin{aligned}
\bar{c}\gamma_\mu\gamma_5 b = \;& C_1^{(A)}\left[\frac{m_b}{\mu}, \frac{m_c}{\mu}, \alpha_s(\mu), w\right]\bar{c}_{v'}\gamma_\mu\gamma_5 b_v \\
&+ C_2^{(A)}\left[\frac{m_b}{\mu}, \frac{m_c}{\mu}, \alpha_s(\mu), w\right]\bar{c}_{v'}v_\mu\gamma_5 b_v \\
&+ C_3^{(A)}\left[\frac{m_b}{\mu}, \frac{m_c}{\mu}, \alpha_s(\mu), w\right]\bar{c}_{v'}v'_\mu\gamma_5 b_v.
\end{aligned} \tag{3.87}$$

The right-hand side contains all dimension three operators with the same quantum numbers as the left-hand side. Higher dimension operators give effects suppressed by powers of $(\Lambda_{\text{QCD}}/m_{c,b})$ and will be considered in the next chapter. In the matching condition of Eqs. (3.86) and (3.87) the transition to HQET is made simultaneously for both quarks. Usually one chooses a subtraction point, $\mu = \bar{m} = \sqrt{m_b m_c}$, which is between the bottom and charm quark masses for the initial value for the C_j's and then runs down to a lower value of μ by using the HQET renormalization group equation. At order α_s, the matching condition contains terms of the order of $\alpha_s(\bar{m})\ln(m_b/m_c)$, but since this logarithm is not very large there is no need to sum all terms of the order of $\alpha_s(\bar{m})^n \ln^n(m_c/m_b)$. Tree-level matching at $\bar{m}$ gives

$$\begin{aligned} C_1^{(V,A)}\left[\frac{m_b}{\bar{m}}, \frac{m_c}{\bar{m}}, \alpha_s(\bar{m}), w\right] &= 1 + \mathcal{O}\left[\alpha_s(\bar{m})\right], \\ C_2^{(V,A)}\left[\frac{m_b}{\bar{m}}, \frac{m_c}{\bar{m}}, \alpha_s(\bar{m}), w\right] &= 0 + \mathcal{O}\left[\alpha_s(\bar{m})\right], \\ C_3^{(V,A)}\left[\frac{m_b}{\bar{m}}, \frac{m_c}{\bar{m}}, \alpha_s(\bar{m}), w\right] &= 0 + \mathcal{O}\left[\alpha_s(\bar{m})\right]. \end{aligned} \tag{3.88}$$

The additional operators $\bar{c}_{v'} v^\mu b_v$ and $\bar{c}_{v'} v'^\mu b_v$ induced at one loop do not cause a loss of predictive power in computing decay rates. In HQET the $\bar{B} \to D^{(*)}$ matrix elements of any operator of the form $\bar{c}_{v'} \Gamma b_v$ (where Γ is a 4×4 matrix in spinor space) can be expressed in terms of the Isgur-Wise function, so the matrix elements of the new operators are related to the matrix elements of the old operators. This was also the case for heavy-light matrix elements in Eq. (3.78).

The calculation of the $C_j^{(V,A)}$ at order α_s is straightforward but somewhat tedious, since these coefficients depend not only on the bottom and charm quark masses but also on the dot product of four velocities $w = v \cdot v'$. In this chapter we shall explicitly calculate the matching condition at the zero-recoil kinematic point, $w = 1$. Here the matching condition simplifies because $\bar{c}_v \gamma_5 b_v = 0$ and $\bar{c}_v \gamma_\mu b_v = \bar{c}_v v_\mu b_v$. Consequently we can write the matching relation as

$$\begin{aligned} \bar{c}\gamma_\mu b &= \eta_V \, \bar{c}_v \gamma_\mu b_v, \\ \bar{c}\gamma_\mu \gamma_5 b &= \eta_A \, \bar{c}_v \gamma_\mu \gamma_5 b_v. \end{aligned} \tag{3.89}$$

As in the case of heavy-light currents, the coefficients η_V and η_A are determined by equating a full QCD matrix element of these currents with the corresponding one in HQET. The matrix element we choose is between an on-shell b-quark state with four-momentum $p_b = m_b v$ and an on-shell c-quark state with four-momentum $p_c = m_c v$. Since $\bar{c}_v \gamma_\mu b_v$ is the conserved current associated with heavy quark flavor symmetry, and $\bar{c}_v \gamma_\mu \gamma_5 b_v$ is related to it by heavy quark spin symmetry, we know the matrix elements of these currents. To all orders in the strong coupling,

$$\left\langle c(v, s')\right| \bar{c}_v \Gamma b_v \left|b(v, s)\right\rangle = \bar{u}(v, s')\, \Gamma\, u(v, s), \tag{3.90}$$

where Γ is any matrix in spinor space (including γ_μ or $\gamma_\mu \gamma_5$), and the right-hand side is absolutely normalized by heavy quark symmetry. This relation is subtraction-point independent and so $\eta_{(V,A)}$ must be μ independent:

$$\mu \frac{\mathrm{d}}{\mathrm{d}\mu} \eta_{(V,A)} = 0. \tag{3.91}$$

The matching condition will be computed by using the procedure outlined at the end of the previous section, so only the finite parts of dimensionally regulated graphs will be computed. The vector current matrix element in QCD is

$$\begin{aligned}&\langle c(p_c, s')|\bar{c}\gamma^\lambda b|b(p_b, s)\rangle\\&\quad = \bar{u}(p_c, s')\left\{1 + \frac{1}{2}\left[R_1^{(c)} + R_1^{(b)}\right]\alpha_s(\mu) + V_1\alpha_s(\mu)\right\}\gamma^\lambda u(p_b, s) + \cdots,\end{aligned} \tag{3.92}$$

where $p_c = m_c v$, $p_b = m_b v$, and the ellipsis denotes terms higher order in α_s. Here $R_1^{(Q)}$ has already been computed, so it only remains to compute the one-particle irreducible vertex at the order of α_s. It is given by the Feynman diagram in Fig. (1.4). In the Feynman gauge Fig. 1.4 yields

$$-ig^2\mu^\varepsilon\left(\frac{4}{3}\right)\int\frac{d^n q}{(2\pi)^n}\frac{\gamma_\alpha(\not{q} + \not{p}_c + m_c)\gamma^\lambda(\not{q} + \not{p}_b + m_b)\gamma^\alpha}{(q^2 + 2p_c\cdot q)(q^2 + 2p_b\cdot q)q^2}. \tag{3.93}$$

The charm and bottom quarks have the same four velocity and so a factor of $\not{p}_{c,b}$ on the far left or right can be replaced by $m_{c,b}$. Hence Eq. (3.93) can be written as

$$\begin{aligned}&-ig^2\mu^\varepsilon\left(\frac{4}{3}\right)\int\frac{d^n q}{(2\pi)^n}\frac{(2m_c v_\alpha + \gamma_\alpha\not{q})\gamma^\lambda(2m_b v^\alpha + \not{q}\gamma^\alpha)}{(q^2 + 2q\cdot p_c)(q^2 + 2q\cdot p_b)q^2}\\&\quad = -ig^2\mu^\varepsilon\left(\frac{4}{3}\right)\int\frac{d^n q}{(2\pi)^n}\\&\qquad\times\left[\frac{4m_c m_b\gamma^\lambda + 2m_c\gamma^\lambda\not{q} + 2m_b\not{q}\gamma^\lambda + (2-n)\not{q}\gamma^\lambda\not{q}}{(q^2 + 2q\cdot p_c)(q^2 + 2q\cdot p_b)q^2}\right].\end{aligned} \tag{3.94}$$

It is convenient to first combine the two quark propagator denominators using the Feynman parameter x, and then combine the result with the gluon propagator using y. Shifting the q integration variable, $q \to q - y[m_c x + m_b(1-x)]v$ and performing the $\mathrm{d}^n q$ integration gives

$$\begin{aligned}&\frac{g^2}{12\pi^2}\gamma^\lambda(4\pi\mu^2)^{\epsilon/2}\int_0^1 dx\int_0^1 y\,dy\,\left(m_x^2 y^2\right)^{-\epsilon/2}\left\{\frac{1}{2}(2-\epsilon)^2\,\Gamma(\epsilon/2)\right.\\&\quad\left.-\Gamma(1+\epsilon/2)\left[\frac{4m_c m_b}{m_x^2 y^2} - 2\frac{m_c + m_b}{m_x y} - (2-\epsilon)\right]\right\}\end{aligned} \tag{3.95}$$

where

$$m_x = m_c x + m_b(1-x).$$

Evaluating the y integral, expanding in ϵ, and rescaling μ to the $\overline{\text{MS}}$ scheme yields

$$\frac{g^2}{6\pi^2}\gamma^\lambda\int_0^1 dx\left[\left(1 + \frac{2m_b m_c}{m_x^2}\right)\frac{1}{\epsilon} + \frac{m_b + m_c}{m_x} - \left(1 + \frac{2m_b m_c}{m_x^2}\right)\ln\left(\frac{m_x}{\mu}\right)\right]. \tag{3.96}$$

Integrating with respect to x and keeping the finite part gives

$$V_1\alpha_s = -\frac{g^2}{6\pi^2}\left[1 + 3\frac{m_b\ln(m_c/\mu) - m_c\ln(m_b/\mu)}{m_b - m_c}\right]. \tag{3.97}$$

Equations (3.90) and (3.92) imply that the matching coefficient is

$$\eta_V = 1 + \alpha_s(\mu)\left[\frac{R_1^{(b)}}{2} + \frac{R_1^{(c)}}{2} + V_1\right] + \cdots, \tag{3.98}$$

where the ellipsis denotes terms of the order of α_s^2 and higher. Using Eq. (3.97) and the finite part of Eq. (3.60), we find that at order α_s,

$$\eta_V = 1 + \frac{\alpha_s(\mu)}{\pi}\left[-2 + \left(\frac{m_b + m_c}{m_b - m_c}\right)\ln\left(\frac{m_b}{m_c}\right)\right]. \tag{3.99}$$

Note that the coefficient of $\alpha_s(\mu)$ is independent of μ. This is a consequence of Eq. (3.91), which states that η_V is independent of the subtraction point μ. Terms higher order in α_s compensate for the dependence of α_s on μ in Eq. (3.99). Usually for numerical evaluation of $\eta_{(V,A)}$ one uses $\mu = \sqrt{m_b m_c} = \bar{m}$.

In the case $m_b = m_c$, the vector current $\bar{c}\gamma^\lambda b$ is a conserved current in QCD and its on-shell matrix element is $\langle c(p_c, s')|\bar{c}\gamma^\lambda b|b(p_b, s)\rangle = \bar{u}(p_c, s')\gamma^\lambda\, u(p_b, s')$, to all orders in α_s. Consequently the coefficient of α_s in Eq. (3.99) vanishes in the limit $m_b = m_c$.

The axial current matching condition is almost the same as in the vector case. In the calculation of the one-particle irreducible vertex, Eq. (3.94) is replaced by

$$\begin{aligned} &-ig^2\mu^\varepsilon\left(\frac{4}{3}\right)\int\frac{d^nq}{(2\pi)^n}\frac{1}{(q^2+2q\cdot p_c)(q^2+2q\cdot p_b)q^2} \\ &\quad\times\left[4m_c m_b\gamma^\lambda\gamma_5 + 2m_c\gamma^\lambda\gamma_5\not{q} + 2m_b\not{q}\gamma^\lambda\gamma_5 + (2-n)\not{q}\gamma^\lambda\not{q}\gamma_5\right]. \end{aligned} \tag{3.100}$$

One can then combine denominators and change the integration variable as for the computation of η_V. The only difference between η_V and η_A is that for η_V, $(2-n)\not{q}\gamma^\lambda\not{q}$ generates the term $(2-n)m_x^2y^2\gamma^\lambda$ on shifting the integration variable, whereas for η_A, $(2-n)\not{q}\gamma^\lambda\not{q}\gamma_5$ generates $-(2-n)m_x^2y^2\gamma^\lambda\gamma_5$. Thus

$$\begin{aligned} \eta_A &= \eta_V + ig^2\left(\frac{4}{3}\right)2(2-n)\int_0^1 dx\int_0^1 2y\,dy\int\frac{d^nq}{(2\pi)^n}\frac{m_x^2y^2}{(q^2-m_x^2y^2)^3} \\ &= \eta_V - \frac{2}{3\pi}\alpha_s(\mu) \\ &= 1 + \frac{\alpha_s(\mu)}{\pi}\left[-\frac{8}{3} + \frac{(m_b+m_c)}{(m_b-m_c)}\ln\left(\frac{m_b}{m_c}\right)\right]. \end{aligned} \tag{3.101}$$

Here $\eta_{(V,A)}$ are important for the $B \to D^{(*)}e\bar{\nu}_e$ differential decay rates near

$w = v \cdot v' = 1$, i.e., $\mathcal{F}_{D^*}(1) = \eta_A$ and $\mathcal{F}_D(1) = \eta_V$ up to corrections suppressed by powers of m_Q.

3.5 Problems

1. The effective Hamiltonian for $B^0 - \bar{B}^0$ mixing is proportional to the operator

$$(\bar{d}\gamma_\mu P_L b)(\bar{d}\gamma^\mu P_L b).$$

After the transition to HQET, it becomes

$$O^{\Delta S=2} = (\bar{d}\gamma_\mu P_L b_v)(\bar{d}\gamma^\mu P_L b_v).$$

Calculate the anomalous dimension of $O^{\Delta S=2}$ at one loop.

2. Analytic expressions for the matching coefficients $C_j^{(V)}$ and $C_j^{(A)}$ can be found in an expansion about $w = 1$.

(a) Show that if the c and b quarks are matched onto the HQET fields $c_{v'}$ and b_v at the common scale $\mu = \bar{m} = \sqrt{m_c m_b}$, then $C_j^{V,A}(w) = 1 + (\alpha_s(\bar{m})/\pi)\delta C_j^{V,A}(w)$, where

$$\delta C_1^{(V)}(1) = -\frac{4}{3} - \frac{1+z}{1-z},$$

$$\delta C_2^{(V)}(1) = -\frac{2(1-z+z\ln z)}{3(1-z)^2},$$

$$\delta C_3^{(V)}(1) = \frac{2z(1-z+\ln z)}{3(1-z)^2},$$

$$\delta C_1^{(A)}(1) = -\frac{8}{3} - \frac{1+z}{1-z}\ln z,$$

$$\delta C_2^{(A)}(1) = -\frac{2[3-2z-z^2+(5-z)z\ln z]}{3(1-z)^3},$$

$$\delta C_3^{(A)}(1) = \frac{2z[1+2z-3z^2+(5z-1)\ln z]}{3(1-z)^3},$$

where $z = m_c/m_b$.

(b) Show that

$$\delta C_1'^{(V)}(1) = -\frac{2[13-9z+9z^2-13z^3+3(2+3z+3z^2+2z^3)\ln z]}{27(1-z)^3},$$

$$\delta C_2'^{(V)}(1) = \frac{2(2+3z-6z^2+z^3+6z\ln z)}{9(1-z)^4},$$

$$\delta C_3'^{(V)}(1) = \frac{2z(1-6z+3z^2+2z^3-6z^2\ln z)}{9(1-z)^4},$$

$$\delta C_1'^{(A)}(1) = -\frac{2[7+9z-9z^2-7z^3+3(2+3z+3z^2+2z^3)\ln z]}{27(1-z)^3},$$

$$\delta C_2'^{(A)}(1) = \frac{2[2-33z+9z^2+25z^3-3z^4-6z(1+7z)\ln z]}{9(1-z)^5},$$

$$\delta C_3'^{(A)}(1) = -\frac{2z[3-25z-9z^2+33z^3-2z^4-6z^2(7+z)\ln z]}{9(1-z)^5},$$

where $'$ denotes differentiation with respect to w.

(c) Using $m_c = 1.4$ GeV and $m_b = 4.8$ GeV, calculate the perturbative QCD corrections to the ratios of form factors $R_1(1)$ and $R_2(1)$ defined in Chapter 2.

3. Prove the identity in Eq. (3.6).

4. Calculate the renormalization of the operators

$$\begin{aligned} O_1 &= \bar{c}_{v'}\, \Gamma i D_\mu\, b_v \\ O_2 &= \bar{c}_{v'}\, \Gamma i \overleftarrow{D}_\mu\, b_v \\ O_3 &= \bar{c}_{v'}\, \Gamma i(v' \cdot D)\, b_v\, v_\mu \\ O_4 &= \bar{c}_{v'}\, \Gamma i(v' \cdot D)\, b_v\, v'_\mu \\ O_5 &= \bar{c}_{v'}\, \Gamma i(v \cdot \overleftarrow{D})\, b_v\, v_\mu \\ O_6 &= \bar{c}_{v'}\, \Gamma i(v \cdot \overleftarrow{D})\, b_v\, v'_\mu \end{aligned}$$

and use it to compute the anomalous dimension matrix for $O_1 - O_6$.

5. Consider the ratio $r_f(w) = f_2(w)/f_1(w)$ of the form factors for $\Lambda_b \to \Lambda_c e \bar{\nu}_e$ decay. Show that in the $m_b \to \infty$ limit, the perturbative α_s correction gives

$$r_f(w) = -\frac{2\alpha_s(m_c)}{3\pi} r(w),$$

where $r(w)$ is defined in Eq. (3.33).

3.6 References

The renormalization Z factors of QCD in the $m_Q \to \infty$ limit, and their implications for heavy quark physics, were discussed in:

M.A. Shifman and M.B. Voloshin, Sov. J. Nucl. Phys. 45 (1987) 292 [Yad. Fiz. 45 (1987) 463]

see also:

H.D. Politzer and M.B. Wise, Phys. Lett. B206 (1988) 681, B208 (1988) 504

Matching of QCD onto HQET was developed in:

A.F. Falk, H. Georgi, B. Grinstein, and M.B. Wise, Nucl. Phys. B343 (1990) 1

Matching of heavy-heavy currents was discussed in:

A.F. Falk and B. Grinstein, Phys. Lett. B247 (1990) 406
M. Neubert, Phys. Rev. D46 (1992) 2212

For some work on two-loop matching and anomalous dimensions, see:

X. Ji and M.J. Musolf, Phys. Lett. B257 (1991) 409
D.J. Broadhurst and A.G. Grozin, Phys. Lett. B267 (1991) 105, Phys. Rev. D52 (1995) 4082
W. Kilian, P. Manakos, and T. Mannel, Phys. Rev. D48 (1993) 1321
M. Neubert, Phys. Lett. B341 (1995) 367
A. Czarnecki, Phys. Rev. Lett. 76 (1996) 4124
G. Amoros, M. Beneke, and M. Neubert, Phys. Lett. B401 (1997) 81

see also:

G.P. Korchemsky and A.V. Radyushkin, Nucl. Phys. B283 (1987) 343, Phys. Lett. B279 (1992) 359

Perturbative matching for heavy-light currents at order $1/m_Q$ is considered in:

A.F. Falk and B. Grinstein, Phys. Lett. B247 (1990) 406
M. Neubert, Phys. Rev. D46 (1992) 1076

For an excellent review on perturbative corrections, see:

M. Neubert, Phys. Rep. 245 (1994) 259

4
Nonperturbative corrections

The effective Lagrangian for heavy quarks has an expansion in powers of $\alpha_s(m_Q)$ and $1/m_Q$. The α_s corrections were discussed in the previous chapter; the $1/m_Q$ corrections are discussed here. By dimensional analysis, these corrections are proportional to $\Lambda_{\rm QCD}/m_Q$, necessarily involve the hadronic scale $\Lambda_{\rm QCD}$, and are nonperturbative in origin. By using the effective Lagrangian approach, we can systematically include these nonperturbative corrections in computations involving hadrons containing a heavy quark.

4.1 The $1/m_Q$ expansion

The HQET Lagrangian including $1/m_Q$ corrections can be derived from the QCD Lagrangian following the procedure of Sec. 2.6. Substituting Eq. (2.43) into the QCD Lagrangian gives

$$\mathcal{L} = \bar{Q}_v\,(iv\cdot D)\,Q_v - \bar{\mathfrak{Q}}_v(iv\cdot D + 2m_Q)\mathfrak{Q}_v + \bar{Q}_v i\,\not{D}\mathfrak{Q}_v + \bar{\mathfrak{Q}}_v i\,\not{D}\,Q_v, \qquad (4.1)$$

using $\not{v}\,Q_v = Q_v$ and $\not{v}\mathfrak{Q}_v = -\mathfrak{Q}_v$. It is convenient to project four vectors into components parallel and perpendicular to the velocity v. The perpendicular component of any four-vector X is defined by

$$X_\perp^\mu \equiv X^\mu - X\cdot v v^\mu. \qquad (4.2)$$

The $i\not{D}$ factors in Eq. (4.1) can be replaced by $i\,\not{D}_\perp$ since $\bar{Q}_v\not{v}\mathfrak{Q}_v = 0$.

The field $\mathfrak{Q}_v$ corresponds to an excitation with mass $2m_Q$, which is the energy required to create a heavy quark–antiquark pair. Here $\mathfrak{Q}_v$ can be integrated out of the theory for physical situations where the use of HQET is justified. This can be done at tree level by solving the $\mathfrak{Q}_v$ equation of motion,

$$(iv\cdot D + 2m_Q)\mathfrak{Q}_v = i\,\not{D}_\perp Q_v, \qquad (4.3)$$

and substituting back into the Lagrangian Eq. (4.1), to give

$$\mathcal{L} = \bar{Q}_v\left(iv\cdot D + i\not{D}_\perp \frac{1}{2m_Q + iv\cdot D} i\not{D}_\perp\right)Q_v$$
$$= \bar{Q}_v\left(iv\cdot D - \frac{1}{2m_Q}\not{D}_\perp\not{D}_\perp\right)Q_v + \cdots, \tag{4.4}$$

where the ellipses denote terms of higher order in the $1/m_Q$ expansion. It is convenient to express the term suppressed by $1/m_Q$ as a sum of two terms, one that violates heavy quark spin symmetry and one that doesn't. Specifically,

$$\not{D}_\perp\not{D}_\perp = \gamma_\mu\gamma_\nu D_\perp^\mu D_\perp^\nu = D_\perp^2 + \frac{1}{2}[\gamma_\mu, \gamma_\nu]D_\perp^\mu D_\perp^\nu. \tag{4.5}$$

Using the identity $[D^\mu, D^\nu] = igG^{\mu\nu}$, and the definition $\sigma_{\mu\nu} = i[\gamma_\mu, \gamma_\nu]/2$, this becomes

$$\not{D}_\perp\not{D}_\perp = D_\perp^2 + \frac{g}{2}\sigma_{\mu\nu}G^{\mu\nu}. \tag{4.6}$$

It is not necessary to include any $\perp$ labels on the μ and ν indices of the $\sigma_{\mu\nu}$ term, since $\bar{Q}_v\sigma_{\mu\nu}v^\mu Q_v = 0$. Substituting Eq. (4.6) into Eq. (4.4) gives

$$\mathcal{L} = \mathcal{L}_0 + \mathcal{L}_1 + \cdots, \tag{4.7}$$

where $\mathcal{L}_0$ is the lowest order Lagrangian Eq. (2.45), and

$$\mathcal{L}_1 = -\bar{Q}_v\frac{D_\perp^2}{2m_Q}Q_v - g\bar{Q}_v\frac{\sigma_{\mu\nu}G^{\mu\nu}}{4m_Q}Q_v. \tag{4.8}$$

In the nonrelativistic constituent quark model, the term $\bar{Q}_v(D_\perp^2/2m_Q)Q_v$ is the heavy quark kinetic energy $\boldsymbol{p}_Q^2/2m_Q$. It breaks heavy quark flavor symmetry because of the explicit dependence on m_Q, but it does not break heavy quark spin symmetry. The magnetic moment interaction term $-g\bar{Q}_v(\sigma_{\mu\nu}G^{\mu\nu}/4m_Q)Q_v$ breaks both heavy quark spin and flavor symmetries.

Equation (4.8) has been derived at tree level. Including loop corrections changes the Lagrangian to

$$\mathcal{L}_1 = -\bar{Q}_v\frac{D_\perp^2}{2m_Q}Q_v - a(\mu)\,g\bar{Q}_v\frac{\sigma_{\mu\nu}G^{\mu\nu}}{4m_Q}Q_v. \tag{4.9}$$

The tree-level matching calculation Eq. (4.8) implies that

$$a(m_Q) = 1 + \mathcal{O}[\alpha_s(m_Q)]. \tag{4.10}$$

The μ dependence of the magnetic moment operator is canceled by the μ dependence of $a(\mu)$. In the leading logarithmic approximation

$$a(\mu) = \left[\frac{\alpha_s(m_Q)}{\alpha_s(\mu)}\right]^{9/(33-2N_q)}, \tag{4.11}$$

where N_q is the number of light quark flavors. Loop effects do not change the coefficient of the heavy quark kinetic energy term. In the next section it is shown that this is a consequence of the reparameterization invariance of the effective Lagrangian.

4.2 Reparameterization invariance

The heavy quark momentum p_Q is given by

$$p_Q = m_Q v + k, \tag{4.12}$$

where v is the heavy quark four velocity and k is its residual momentum. This decomposition of p_Q into v and k is not unique. Typically k is of the order of $\Lambda_{\rm QCD}$, which is much smaller than m_Q. A small change in the four velocity of the order of $\Lambda_{\rm QCD}/m_Q$ can be compensated by a change in the residual momentum:

$$\begin{aligned} v &\to v + \varepsilon/m_Q, \\ k &\to k - \varepsilon. \end{aligned} \tag{4.13}$$

Since the four velocity satisfies $v^2 = 1$, the parameter ε must satisfy

$$v \cdot \varepsilon = 0, \tag{4.14}$$

neglecting terms of order $(\varepsilon/m_Q)^2$. In addition to the changes of v and k in Eqs. (4.13), the heavy quark spinor Q_v must also change to preserve the constraint $\not{v} Q_v = Q_v$. Consequently, if

$$Q_v \to Q_v + \delta Q_v, \tag{4.15}$$

δQ_v satisfies

$$\left(\not{v} + \frac{\not{\varepsilon}}{m_Q}\right)(Q_v + \delta Q_v) = Q_v + \delta Q_v. \tag{4.16}$$

At linear order in (ε/m_Q), one finds

$$(1 - \not{v})\delta Q_v = \frac{\not{\varepsilon}}{m_Q} Q_v. \tag{4.17}$$

Therefore a suitable choice for the change in Q_v is

$$\delta Q_v = \frac{\not{\varepsilon}}{2m_Q} Q_v. \tag{4.18}$$

This satisfies $\not{v}\delta Q_v = -\delta Q_v$, since $v \cdot \varepsilon = 0$, so that Eq. (4.17) holds. The solution to Eq. (4.17) is not unique, and we have chosen one that preserves the normalization of the $i v \cdot D$ term. Other choices are equivalent to the above by a simple redefinition of the field.

In summary, the Lagrange density in Eq. (4.7) must be invariant under the combined changes

$$\begin{aligned} v &\to v + \varepsilon/m_Q, \\ Q_v &\to e^{i\varepsilon\cdot x}\left(1 + \frac{\not\varepsilon}{2m_Q}\right)Q_v, \end{aligned} \tag{4.19}$$

where the prefactor $e^{i\varepsilon\cdot x}$ causes a shift in the residual momentum $k \to k - \varepsilon$. Under the transformation in Eq. (4.19),

$$\begin{aligned} \mathcal{L}_0 &\to \mathcal{L}_0 + \frac{1}{m_Q}\bar{Q}_v(i\varepsilon\cdot D)Q_v, \\ \mathcal{L}_1 &\to \mathcal{L}_1 - \frac{1}{m_Q}\bar{Q}_v(i\varepsilon\cdot D)Q_v. \end{aligned} \tag{4.20}$$

Consequently the Lagrangian, $\mathcal{L}_0 + \mathcal{L}_1$, is reparameterization invariant. This would not be the case if the coefficient of the kinetic energy deviated from unity. There can be no corrections to the coefficient of the kinetic energy operator as long as the theory is regularized in a way that preserves reparameterization invariance. Dimensional regularization is such a regulator, since the arguments made in this section hold in n dimensions.

An important feature of reparameterization invariance is that it connects different orders in the $1/m_Q$ expansion, since the transformation Eq. (4.19) explicitly involves m_Q. Thus it can be used to fix the form of some $1/m_Q$ corrections using only information from lower order terms in $1/m_Q$, as was done for the kinetic energy term.

4.3 Masses

Heavy quark symmetry can be used to obtain relations between hadron masses. The hadron mass in the effective theory is $m_H - m_Q$, since the heavy quark mass m_Q has been subtracted from all energies in the field redefinition in Eq. (2.43). At order m_Q, all heavy hadrons containing Q are degenerate, and have the same mass m_Q. At the order of unity, the hadron masses get the contribution

$$\frac{1}{2}\langle H^{(Q)}|\mathcal{H}_0|H^{(Q)}\rangle \equiv \bar{\Lambda}, \tag{4.21}$$

where $\mathcal{H}_0$ is the order $1/m_Q^0$ terms in the HQET Hamiltonian obtained from the Lagrangian term $\bar{Q}_v(iv\cdot D)Q_v$, as well as the terms involving light quarks and gluons. In this section, the hadron states $|H^{(Q)}\rangle$ are in the effective theory with $v = v_r = (1, \mathbf{0})$. The factor $1/2$ arises from the normalization introduced in Sec. 2.7. Here $\bar{\Lambda}$ is a parameter of HQET and has the same value for all particles in a spin-flavor multiplet. The values will be denoted by $\bar{\Lambda}$ for the B, B^*, D, and D^* states, $\bar{\Lambda}_\Lambda$ for the Λ_b and Λ_c, and $\bar{\Lambda}_\Sigma$ for the Σ_b, Σ_b^*, Σ_c, and Σ_c^*. In the $SU(3)$ limit, $\bar{\Lambda}$ does not depend on the light quark flavor. If $SU(3)$ breaking is included, $\bar{\Lambda}$ is different for the $B_{u,d}$ and B_s mesons, and will be denoted by $\bar{\Lambda}_{u,d}$ and $\bar{\Lambda}_s$, respectively.

At order $1/m_Q$, there is an additional contribution to the hadron masses given by the expectation value of the $1/m_Q$ correction to the Hamiltonian:

$$\mathcal{H}_1 = -\mathcal{L}_1 = \bar{Q}_v \frac{D_\perp^2}{2m_Q} Q_v + a(\mu) g \bar{Q}_v \frac{\sigma_{\alpha\beta} G^{\alpha\beta}}{4m_Q} Q_v. \tag{4.22}$$

The matrix elements of the two terms in Eq. (4.22) define two nonperturbative parameters, λ_1 and λ_2:

$$\begin{aligned} 2\lambda_1 &= -\langle H^{(Q)} | \bar{Q}_{v_r} D_\perp^2 Q_{v_r} | H^{(Q)} \rangle, \\ 16(\mathbf{S}_Q \cdot \mathbf{S}_\ell)\lambda_2(m_Q) &= a(\mu) \langle H^{(Q)} | \bar{Q}_{v_r} g \sigma_{\alpha\beta} G^{\alpha\beta} Q_{v_r} | H^{(Q)} \rangle. \end{aligned} \tag{4.23}$$

Here λ_1 is independent of m_Q, and λ_2 depends on m_Q through the logarithmic m_Q dependence of $a(\mu)$ in Eq. (4.11); $\lambda_{1,2}$ have the same value for all states in a given spin-flavor multiplet and are expected to be of the order of $\Lambda_{\rm QCD}^2$. The naive expectation that the heavy quark kinetic energy is positive suggests that λ_1 should be negative. The λ_2 matrix element transforms like $\mathbf{S}_Q \cdot \mathbf{S}_\ell$ under the spin symmetry, since that is the transformation property of $\bar{Q}_{v_r} \sigma_{\alpha\beta} G^{\alpha\beta} Q_{v_r}$. Only the two upper components of Q_{v_r} are nonzero, since $\gamma^0 Q_{v_r} = Q_{v_r}$, and $\bar{Q}_{v_r} \sigma_{\alpha\beta} G^{\alpha\beta} Q_{v_r}$ reduces to the matrix element of $\bar{Q}_{v_r} \boldsymbol{\sigma} \cdot \boldsymbol{B} Q_{v_r}$, where $\boldsymbol{B}$ is the chromomagnetic field. The operator $\bar{Q}_{v_r} \boldsymbol{\sigma} Q_{v_r}$ is the heavy quark spin, and the matrix element of $\boldsymbol{B}$ in the hadron must be proportional to the spin of the light degrees of freedom, by rotational invariance and time-reversal invariance, so that the chromomagnetic operator contribution is proportional to $\mathbf{S}_Q \cdot \mathbf{S}_\ell$. Using $\mathbf{S}_Q \cdot \mathbf{S}_\ell = (\mathbf{J}^2 - \mathbf{S}_Q^2 - \mathbf{S}_\ell^2)/2$, one finds that

$$\begin{aligned} m_B &= m_b + \bar{\Lambda} - \frac{\lambda_1}{2m_b} - \frac{3\lambda_2(m_b)}{2m_b}, \\ m_{B^*} &= m_b + \bar{\Lambda} - \frac{\lambda_1}{2m_b} + \frac{\lambda_2(m_b)}{2m_b}, \\ m_{\Lambda_b} &= m_b + \bar{\Lambda}_\Lambda - \frac{\lambda_{\Lambda,1}}{2m_b}, \\ m_{\Sigma_b} &= m_b + \bar{\Lambda}_\Sigma - \frac{\lambda_{\Sigma,1}}{2m_b} - \frac{2\lambda_{\Sigma,2}(m_b)}{m_b}, \\ m_{\Sigma_b^*} &= m_b + \bar{\Lambda}_\Sigma - \frac{\lambda_{\Sigma,1}}{2m_b} + \frac{\lambda_{\Sigma,2}(m_b)}{m_b}, \\ m_D &= m_c + \bar{\Lambda} - \frac{\lambda_1}{2m_c} - \frac{3\lambda_2(m_c)}{2m_c}, \end{aligned} \tag{4.24}$$

$$
\begin{aligned}
m_{D^*} &= m_c + \bar{\Lambda} - \frac{\lambda_1}{2m_c} + \frac{\lambda_2(m_c)}{2m_c}, \\
m_{\Lambda_c} &= m_c + \bar{\Lambda}_\Lambda - \frac{\lambda_{\Lambda,1}}{2m_c}, \\
m_{\Sigma_c} &= m_c + \bar{\Lambda}_\Sigma - \frac{\lambda_{\Sigma,1}}{2m_c} - \frac{2\lambda_{\Sigma,2}(m_c)}{m_c}, \\
m_{\Sigma_c^*} &= m_c + \bar{\Lambda}_\Sigma - \frac{\lambda_{\Sigma,1}}{2m_c} + \frac{\lambda_{\Sigma,2}(m_c)}{m_c}.
\end{aligned}
$$

The average mass of a heavy quark spin symmetry multiplet, e.g., $(3m_{P^*} + m_P)/4$ for the meson multiplet, does not depend on λ_2. The magnetic interaction λ_2 is responsible for the $B^* - B$ and $D^* - D$ splittings. The observed value of the $B^* - B$ mass difference gives $\lambda_2\,(m_b) \simeq 0.12\ \text{GeV}^2$.

Equations (4.24) give the meson mass relation

$$0.49\ \text{GeV}^2 \simeq m_{B^*}^2 - m_B^2 \simeq 4\lambda_2 \simeq m_{D^*}^2 - m_D^2 \simeq 0.55\ \text{GeV}^2, \qquad (4.25)$$

up to corrections of order $1/m_{b,c}$, and ignoring the weak m_Q dependence of λ_2. Similarly, one finds that

$$
\begin{aligned}
90 \pm 3\ \text{MeV} = m_{B_s} - m_{B_d} &= \bar{\Lambda}_s - \bar{\Lambda}_{u,d} = m_{D_s} - m_{D_d} = 99 \pm 1\ \text{MeV}, \\
345 \pm 9\ \text{MeV} = m_{\Lambda_b} - m_B &= \bar{\Lambda}_\Lambda - \bar{\Lambda}_{u,d} = m_{\Lambda_c} - m_D = 416 \pm 1\ \text{MeV}.
\end{aligned}
\qquad (4.26)
$$

The parameters λ_1 and λ_2 are nonperturbative parameters of QCD and have not been computed from first principles. It might appear that very little has been gained by using Eqs. (4.24) for the hadron masses in terms of $\bar{\Lambda}$, λ_1, and λ_2. However, the same hadronic matrix elements also occur in other quantities, such as form factors and decay rates. One can then use the values of $\bar{\Lambda}$, λ_1, and λ_2 obtained by fitting to the hadron masses to compute the form factors and decay rates, without making any model dependent assumptions. An example of this is given in Problems 2–3.

4.4 $\Lambda_b \to \Lambda_c e \bar{\nu}_e$ decay

The HQET predictions for $\Lambda_b \to \Lambda_c$ form factors were discussed earlier in Sec. 2.11. Recall that the most general form factors are

$$
\begin{aligned}
\langle \Lambda_c(p', s') | \bar{c}\gamma^\nu b | \Lambda_b(p, s) \rangle &= \bar{u}(p', s')[f_1\gamma^\nu + f_2 v^\nu + f_3 v'^\nu] u(p, s), \\
\langle \Lambda_c(p', s') | \bar{c}\gamma^\nu \gamma_5 b | \Lambda_b(p, s) \rangle &= \bar{u}(p', s')[g_1\gamma^\nu + g_2 v^\nu + g_3 v'^\nu] \gamma_5 u(p, s),
\end{aligned}
\qquad (4.27)
$$

where $p' = m_{\Lambda_c} v'$ and $p = m_{\Lambda_b} v$. It is convenient for the HQET analysis to consider the form factors f_j and g_j as functions of the dimensionless variable

$w = v \cdot v'$. Heavy quark symmetry implies that

$$\langle \Lambda_c(v', s')|\bar{c}_{v'}\Gamma b_v|\Lambda_b(v, s)\rangle = \zeta(w)\, \bar{u}(v', s')\Gamma u(v, s), \tag{4.28}$$

with $\zeta(1) = 1$. Consequently the form factors are

$$f_1 = g_1 = \zeta(w), \qquad f_2 = f_3 = g_2 = g_3 = 0. \tag{4.29}$$

In Sec. 3.4 perturbative QCD corrections to the matching of heavy quark currents were computed. For the vector current, new operators of the form $v^\mu \bar{c}_{v'} b_v$ and $v'^\mu \bar{c}_{v'} b_v$ were induced with calculable coefficients. These additional terms do not represent any loss of predictive power because Eq. (4.28) gives the matrix elements of these new operators in terms of the same Isgur-Wise function $\zeta(w)$.

In this section, nonperturbative corrections suppressed by $\Lambda_{\rm QCD}/m_{c,b}$ are considered. These corrections arise from two sources. There are time-ordered products of the $1/m_Q$ terms in the Lagrangian with the heavy quark current. These terms can be thought of as correcting the hadron states in HQET at order $1/m_Q$, or equivalently, as producing a $1/m_Q$ correction to the current, and leaving the states unchanged. For example, the chromomagnetic $1/m_c$ correction to the Lagrangian gives a correction to the current $\bar{c}_{v'}\Gamma b_v$ of

$$-i\frac{a(\mu)}{2}\int d^4x\, T\left(g\bar{c}_{v'}\frac{\sigma^{\mu\nu}G_{\mu\nu}}{2m_c}c_{v'}\Big|_x\ \bar{c}_{v'}\Gamma b_v\Big|_0\right). \tag{4.30}$$

Spin symmetry implies that for $\Lambda_b \to \Lambda_c$ matrix elements in HQET, the above quark–gluon operator is equivalent to the hadronic operator

$$\bar{\Lambda}^{(c)}(v', s')\sigma_{\mu\nu}\frac{(1+\not{v}')}{2}\Gamma \Lambda^{(b)}(v, s)\frac{X^{\mu\nu}}{m_c}, \tag{4.31}$$

where $X_{\mu\nu}$ depends on v and v' and is antisymmetric in its indices μ and ν. The $\sigma_{\mu\nu}$ matrix must be next to $\bar{\Lambda}^{(c)}(v', s')$, and the Γ matrix must be next to $\Lambda^{(b)}(v, s)$ because these matrices were next to $\bar{c}_{v'}$ and b_v in Eq. (4.30). The projector $(1+\not{v}')/2$ arises because $\sigma_{\mu\nu}$ and Γ were multiplied on the right and left by $c_{v'}$ and $\bar{c}_{v'}$, respectively, in Eq. (4.30). The only possibility for X is $X_{\mu\nu} \propto v_\mu v'_\nu - v_\nu v'_\mu$, with the constant of proportionality a function of w. With this form for $X_{\mu\nu}$, Eq. (4.31) is zero since $(1+\not{v}')\sigma^{\mu\nu}(1+\not{v}')v'_\mu = 0$. Thus the chromomagnetic $1/m_c$ correction to the charm quark part of the Lagrangian has no effect on the $\Lambda_b \to \Lambda_c e\bar{\nu}_e$ form factors. Clearly, the same conclusion holds for the $1/m_b$ chromomagnetic correction to the bottom quark part of the Lagrangian.

The kinetic energies of the bottom and charm quarks do not violate heavy quark spin symmetry so they preserve $f_2 = f_3 = g_2 = g_3 = 0$ and can be absorbed into a redefinition of the Isgur-Wise function $\zeta(w)$. It is important to know if this correction to ζ preserves the normalization condition $\zeta(1) = 1$ at zero recoil. One can show that the normalization is preserved by an argument similar to that used in proving the Ademollo-Gatto theorem. The $1/m_Q$ kinetic energy term in the Lagrange density changes the $|\Lambda_Q(v, s)\rangle$ state in HQET to the state $|\Lambda_Q(v, s)\rangle + (\varepsilon/m_Q)|S_Q(v, s)\rangle + \cdots$, where $|S_Q(v, s)\rangle$ is a state orthogonal to $|\Lambda_Q(v, s)\rangle$, ε is of the order of Λ_{QCD}, and the ellipses denote terms suppressed by more powers of $1/m_Q$. At zero recoil, $\bar{c}_v \Gamma b_v$ is a charge of heavy quark spin flavor symmetry so it takes $|\Lambda_b(v, s)\rangle$ to the state $|\Lambda_c(v, s)\rangle$, which is orthogonal to $|S_c(v, s)\rangle$. Consequently at order $1/m_Q$ the heavy quark kinetic energies preserve Eq. (4.29) and do not change the normalization of ζ at zero recoil. Equivalently, one can use an analysis analogous to that for the chromomagnetic operator. The time-ordered product

$$-i \int d^4x T\left(g\bar{c}_{v'} \frac{D_\perp^2}{2m_c} c_{v'} \Big|_x \; \bar{c}_{v'} \Gamma b_v \Big|_0 \right) \tag{4.32}$$

is equivalent to the hadronic operator

$$\bar{\Lambda}^{(c)}(v', s') \frac{(1 + \not{v}')}{2} \Gamma \Lambda^{(b)}(v, s) \frac{\chi_1}{m_c}, \tag{4.33}$$

where χ_1 is an arbitrary function of w. Similarly, the b-quark kinetic energy gives a correction term

$$\bar{\Lambda}^{(c)}(v', s') \Gamma \frac{(1 + \not{v})}{2} \Lambda^{(b)}(v, s) \frac{\chi_1}{m_b}. \tag{4.34}$$

The two χ_1's are the same (see Problem 4), since one can relate the form of the matrix elements of the two possible time-ordered products by $v \leftrightarrow v'$ and $c \leftrightarrow b$. Equations (4.33) and (4.34) give the following correction terms to the form factors:

$$\begin{aligned} \delta f_1 &= \chi_1 \left(\frac{1}{m_c} + \frac{1}{m_b} \right), \\ \delta g_1 &= \chi_1 \left(\frac{1}{m_c} + \frac{1}{m_b} \right), \\ \delta f_2 &= \delta f_3 = \delta g_2 = \delta g_3 = 0. \end{aligned} \tag{4.35}$$

This corresponds to a redefinition of the Isgur-Wise function:

$$\zeta(w) \to \zeta(w) + \chi_1(w) \left(\frac{1}{m_c} + \frac{1}{m_b} \right). \tag{4.36}$$

At zero recoil, for $m_b = m_c$, the vector current matrix element is normalized, since it is a symmetry generator of the full QCD theory. Since $\zeta(1) = 1$, this implies that $\chi_1(1) = 0$. As a result, the effects of χ_1 can be reabsorbed into ζ by the redefinition in Eq. (4.36), without affecting the normalization at zero recoil.

In addition to the $1/m_Q$ corrections to the Lagrange density, there are order $1/m_Q$ terms that correct the relation between currents in full QCD and HQET.

These terms arise when one includes the $1/m_Q$ corrections to the relation between the quark fields in QCD and HQET. At tree level,

$$Q = e^{-im_Q v\cdot x}\left(1 + i\frac{\not{D}}{2m_Q}\right)Q_v, \tag{4.37}$$

where the relation in Eq. (2.43) and the solution for $\mathfrak{Q}_v$ in Eq. (4.3) have been used. One could equally well have a $\perp$ subscript on the covariant derivative. These two forms for Eq. (4.37) are equivalent, since the difference vanishes by the equation of motion $(v\cdot D)Q_v = 0$. Using Eq. (4.37) the relation between the QCD current and HQET operators to order $1/m_Q$ is

$$\begin{aligned}\bar{c}\gamma^\nu b &= \bar{c}_{v'}\left(\gamma^\nu - \frac{i\overleftarrow{D}_\mu}{2m_c}\gamma^\mu\gamma^\nu + \gamma^\nu\gamma^\mu\frac{iD_\mu}{2m_b}\right)b_v,\\ \bar{c}\gamma^\nu\gamma_5 b &= \bar{c}_{v'}\left(\gamma^\nu\gamma_5 - \frac{i\overleftarrow{D}_\mu}{2m_c}\gamma^\mu\gamma^\nu\gamma_5 + \gamma^\nu\gamma_5\gamma^\mu\frac{iD_\mu}{2m_b}\right)b_v.\end{aligned} \tag{4.38}$$

Heavy quark spin symmetry implies for $\Lambda_b \to \Lambda_c$ matrix elements in HQET, one can use

$$\bar{c}_{v'}i\overleftarrow{D}_\mu\Gamma b_v = \bar{\Lambda}^{(c)}(v', s')\Gamma\Lambda^{(b)}(v, s)[Av_\mu + Bv'_\mu], \tag{4.39}$$

where A and B are functions of w. The equation of motion $(iv'\cdot D)c_{v'} = 0$ implies that contracting v'^μ into the above give zero, so

$$B = -Aw. \tag{4.40}$$

The function A can be expressed in terms of $\bar{\Lambda}_\Lambda$ and the Isgur-Wise function ζ. To show this note that

$$\begin{aligned}&\langle\Lambda_c(v', s')|i\partial_\mu(\bar{c}_{v'}\Gamma b_v)|\Lambda_b(v, s)\rangle\\ &\quad= \left[(m_{\Lambda_b} - m_b)v_\mu - (m_{\Lambda_c} - m_c)v'_\mu\right]\langle\Lambda_c(v', s')|\bar{c}_{v'}\Gamma b_v|\Lambda_b(v, s)\rangle\\ &\quad= \bar{\Lambda}_\Lambda(v - v')_\mu\zeta\bar{u}(v', s')\Gamma u(v, s).\end{aligned} \tag{4.41}$$

So for $\Lambda_b \to \Lambda_c$ matrix elements in HQET,

$$\begin{aligned}i\partial_\mu(\bar{c}_{v'}\Gamma b_v) &= \bar{c}_{v'}i\overleftarrow{D}_\mu\Gamma b_v + \bar{c}_{v'}\Gamma iD_\mu b_v\\ &= \bar{\Lambda}_\Lambda(v - v')_\mu\zeta\bar{\Lambda}^{(c)}(v', s')\Gamma\Lambda^{(b)}(v, s).\end{aligned} \tag{4.42}$$

Contracting v^μ into this and using the equation of motion $(iv^\mu D_\mu)b_v = 0$ implies that

$$A(1 - w^2) = \bar{\Lambda}_\Lambda\zeta(1 - w), \tag{4.43}$$

giving

$$A = \frac{\bar{\Lambda}_\Lambda \zeta(w)}{1+w}. \tag{4.44}$$

In summary, putting all the pieces together gives

$$\bar{c}_{v'} i\overleftarrow{D}_\mu \Gamma b_v = \frac{\bar{\Lambda}_\Lambda \zeta}{1+w}\, \bar{\Lambda}^{(c)}(v', s')\Gamma\Lambda^{(b)}(v, s)(v_\mu - w v'_\mu). \tag{4.45}$$

For the operator with the derivative on the bottom quark, one uses

$$\begin{aligned} \bar{c}_{v'} \Gamma i D_\mu b_v &= -(\bar{b}_v i\overleftarrow{D}_\mu \bar{\Gamma} c_{v'})^\dagger \\ &= -\frac{\bar{\Lambda}_\Lambda \zeta}{1+w}\, \bar{\Lambda}^{(c)}(v', s')\Gamma\Lambda^{(b)}(v, s)(v'_\mu - w v_\mu). \end{aligned} \tag{4.46}$$

Using these results with $\Gamma = \gamma^\mu \gamma^\nu$, and so on, together with the fact that the effect of $1/m_Q$ corrections to the Lagrangian can be absorbed into a redefinition of ζ, yields the following expression for the $\Lambda_b \to \Lambda_c e \bar{\nu}_e$ form factors at order $1/m_Q$:

$$\begin{aligned} f_1 &= \left[1 + \left(\frac{\bar{\Lambda}_\Lambda}{2m_c} + \frac{\bar{\Lambda}_\Lambda}{2m_b}\right)\right]\zeta(w), \\ f_2 &= -\frac{\bar{\Lambda}_\Lambda}{m_c}\left(\frac{1}{1+w}\right)\zeta(w), \\ f_3 &= -\frac{\bar{\Lambda}_\Lambda}{m_b}\left(\frac{1}{1+w}\right)\zeta(w), \\ g_1 &= \left[1 - \left(\frac{\bar{\Lambda}_\Lambda}{2m_c} + \frac{\bar{\Lambda}_\Lambda}{2m_b}\right)\left(\frac{1-w}{1+w}\right)\right]\zeta(w), \\ g_2 &= -\frac{\bar{\Lambda}_\Lambda}{m_c}\left(\frac{1}{1+w}\right)\zeta(w), \\ g_3 &= \frac{\bar{\Lambda}_\Lambda}{m_b}\left(\frac{1}{1+w}\right)\zeta(w). \end{aligned} \tag{4.47}$$

The leading order predictions for the form factors in Eq. (4.29) involved a single unknown function $\zeta(w)$. The result including $1/m_Q$ corrections involves a single unknown function, as well as the nonperturbative constant $\bar{\Lambda}_\Lambda$. Many of the leading order relations survive even when the $1/m_Q$ corrections are included in Λ_b decay form factors. In the next section, we will see that fewer relations hold for meson decay including $1/m_Q$ corrections, but some important ones continue to hold even at this order. The Λ_{QCD}/m_Q corrections are expected to be numerically small, of the order of ~10–20%.

At the zero-recoil point $w = 1$, the matrix elements of the vector and axial vector currents in Λ_b decay become

$$\begin{aligned}\langle \Lambda_c(p', s')|\bar{c}\gamma^\nu b|\Lambda_b(p, s)\rangle &= [f_1 + f_2 + f_3]v^\nu\, u(p', s')u(p, s),\\ \langle \Lambda_c(p', s')|\bar{c}\gamma^\nu\gamma_5 b|\Lambda_b(p, s)\rangle &= g_1\, \bar{u}(p', s')\gamma^\nu\gamma_5 u(p, s).\end{aligned} \tag{4.48}$$

One can see from Eq. (4.47) that at $w = 1$, $f_1 + f_2 + f_3$ and g_1 do not receive any nonperturbative $1/m_Q$ corrections, so that the decay matrix element has no $1/m_Q$ corrections at zero recoil, a result known as Luke's theorem. Note that the individual form factors can have $1/m_Q$ corrections at zero recoil, but the matrix element does not. A similar result will be proven for B decays in the next section.

4.5 $\bar{B} \to D^{(*)}e\bar{\nu}_e$ decay and Luke's theorem

The analysis of $1/m_Q$ corrections for $\Lambda_b \to \Lambda_c$ semileptonic decay can be repeated for $\bar{B} \to D^{(*)}$ semileptonic decay. To determine the $1/m_Q$ corrections using the weak currents in Eq. (4.38), one needs the matrix elements of $\bar{c}_{v'}i\overleftarrow{D}_\mu\Gamma b_v$ and $\bar{c}_{v'}\Gamma iD_\mu b_v$ between $\bar{B}$ and $D^{(*)}$ meson states at leading order in $1/m_Q$. For this, one can use

$$\begin{aligned}\bar{c}_{v'}i\overleftarrow{D}_\mu\Gamma b_v &= \mathrm{Tr}\, \bar{H}_{v'}^{(c)}\Gamma H_v^{(b)} M_\mu(v, v')\\ \bar{c}_{v'}\Gamma iD_\mu b_v &= -(\bar{b}_v i\overleftarrow{D}_\mu\bar{\Gamma}c_{v'})^\dagger = -\mathrm{Tr}\, \bar{H}_{v'}^{(c)}\Gamma H_v^{(b)}\bar{M}_\mu(v', v)\end{aligned} \tag{4.49}$$

where

$$M_\mu(v, v') = \xi_+(v + v')_\mu + \xi_-(v - v')_\mu - \xi_3\gamma_\mu \tag{4.50}$$

is the most general bispinor constructed out of v and v'. There is no term proportional to $\epsilon_{\mu\alpha\beta\nu}v^\alpha v'^\beta\gamma^\nu\gamma_5$ since it can be eliminated by using the three-γ matrix identity in Eq. (1.119) to write

$$-i\epsilon_{\mu\alpha\beta\nu}v^\alpha v'^\beta\gamma^\nu\gamma_5 = \gamma_\mu \not{v}\not{v}' - v_\mu\not{v}' - w\gamma_\mu + v'_\mu\not{v}, \tag{4.51}$$

which can be absorbed into the other terms using $H_v^{(b)}\not{v} = -H_v^{(b)}$, $\not{v}'\bar{H}_{v'}^{(c)} = -\bar{H}_{v'}^{(c)}$. The equation of motion, $(iv'\cdot D)c_{v'} = 0$, implies that

$$\xi_+(w + 1) - \xi_-(w - 1) + \xi_3 = 0. \tag{4.52}$$

By an argument similar to that used to derive Eq. (4.41), one finds that for $\bar{B} \to D^{(*)}$ matrix elements,

$$\begin{aligned}i\partial_\mu(c_{v'}\Gamma b_v) &= \bar{c}_{v'}i\overleftarrow{D}_\mu\Gamma b_v + c_{v'}\Gamma iD_\mu b_v\\ &= -\bar{\Lambda}(v - v')_\mu\, \xi\, \mathrm{Tr}\, \bar{H}_{v'}^{(c)}\Gamma H_v^{(b)},\end{aligned} \tag{4.53}$$

which implies using Eqs. (4.49) and (4.50) that

$$\xi_-(w) = \frac{1}{2}\bar{\Lambda}\xi(w). \tag{4.54}$$

When combined with Eq. (4.52), this yields

$$\xi_+(w) = \frac{w-1}{2(w+1)}\bar{\Lambda}\xi(w) - \frac{\xi_3(w)}{w+1}. \tag{4.55}$$

The $1/m_Q$ corrections to the $\bar{B} \to D^{(*)}$ form factors that were defined in Eq. (2.84) from the $1/m_Q$ terms in the currents given in Eq. (4.38) are

$$\begin{aligned}
\delta h_+ &= [(1+w)\xi_+ + \xi_3]\left(\frac{1}{2m_c} + \frac{1}{2m_b}\right) - (w-1)\xi_-\left(\frac{1}{2m_c} + \frac{1}{2m_b}\right),\\
\delta h_- &= [(1+w)\xi_+ + 3\xi_3]\left(\frac{1}{2m_c} - \frac{1}{2m_b}\right) - (w+1)\xi_-\left(\frac{1}{2m_c} - \frac{1}{2m_b}\right),\\
\delta h_V &= \xi_-\left(\frac{1}{m_c} + \frac{1}{m_b}\right) - \xi_3\left(\frac{1}{m_b}\right),\\
\delta h_{A_1} &= \xi_+\left(\frac{1}{m_c} + \frac{1}{m_b}\right) + \frac{\xi_3}{1+w}\left(\frac{1}{m_c} + \frac{2-w}{m_b}\right),\\
\delta h_{A_2} &= (\xi_+ - \xi_-)\left(\frac{1}{m_c}\right),\\
\delta h_{A_3} &= -\xi_3\left(\frac{1}{m_b}\right) + \xi_-\left(\frac{1}{m_b}\right) + \xi_+\left(\frac{1}{m_c}\right),
\end{aligned} \tag{4.56}$$

where ξ_+ and ξ_- are given in Eqs. (4.54) and (4.55).

One also needs to evaluate the $1/m_Q$ corrections from the Lagrangian. The time-ordered product of the c-quark chromomagnetic operator with the weak currents, Eq. (4.30), can be written as

$$\mathrm{Tr}\, \bar{H}_{v'}^{(c)} \sigma_{\mu\nu} \frac{(1+\not{v}')}{2} \Gamma H_v^{(b)} \frac{X^{\mu\nu}}{2m_c}, \tag{4.57}$$

as for the $\Lambda_b \to \Lambda_c$ case. The only difference is that $X_{\mu\nu}$ is now a general bispinor that is antisymmetric in μ and ν. The most general form for $X_{\mu\nu}$ that does not give a vanishing contribution is

$$X_{\mu\nu} = i\chi_2(v_\mu\gamma_\nu - v_\nu\gamma_\mu) - 2\chi_3\sigma_{\mu\nu}. \tag{4.58}$$

A similar result holds for the b-quark chromomagnetic moment. The c-quark kinetic energy term gives a time-ordered product contribution

$$-\mathrm{Tr}\, \bar{H}_{v'}^{(c)} \frac{(1+\not{v}')}{2} \Gamma H_v^{(b)} \frac{\chi_1}{m_c}, \tag{4.59}$$

with a similar expression for the b-quark kinetic energy. These give

$$\begin{aligned}
\delta h_+ &= \chi_1\left(\frac{1}{m_c}+\frac{1}{m_b}\right) - 2(w-1)\chi_2\left(\frac{1}{m_c}+\frac{1}{m_b}\right) + 6\chi_3\left(\frac{1}{m_c}+\frac{1}{m_b}\right),\\
\delta h_- &= 0,\\
\delta h_V &= \chi_1\left(\frac{1}{m_c}+\frac{1}{m_b}\right) - 2(w-1)\chi_2\left(\frac{1}{m_b}\right) - 2\chi_3\left(\frac{1}{m_c}-\frac{3}{m_b}\right),\\
\delta h_{A_1} &= \chi_1\left(\frac{1}{m_c}+\frac{1}{m_b}\right) - 2(w-1)\chi_2\left(\frac{1}{m_b}\right) - 2\chi_3\left(\frac{1}{m_c}-\frac{3}{m_b}\right),\\
\delta h_{A_2} &= 2\chi_2\left(\frac{1}{m_c}\right),\\
\delta h_{A_3} &= \chi_1\left(\frac{1}{m_c}+\frac{1}{m_b}\right) - 2\chi_3\left(\frac{1}{m_c}-\frac{3}{m_b}\right) - 2\chi_2\left(\frac{1}{m_c}+\frac{w-1}{m_b}\right).
\end{aligned} \tag{4.60}$$

The expressions for the form factors are given by adding Eqs. (4.56) and (4.60) to Eq. (2.95). In addition, there are the perturbative corrections discussed in Chapter 3. We will see in the next section that there is a connection between these two seemingly very different kinds of terms.

The $1/m_Q$ corrections to the form factors are parameterized in terms of one unknown constant $\bar{\Lambda}$, and four unknown functions ξ_3, χ_{1-3}, so there are several new functions in the expressions for the meson decay form factors at order $1/m_Q$. At zero recoil, the $\bar{B} \to D$ matrix element of the vector current is normalized when $m_c = m_b$. This gives the constraint $\chi_1(1) + 6\chi_3(1) = 0$. There is also a constraint from the $\bar{B}^* \to D^*$ matrix element being absolutely normalized at $w = 1$ when $m_b = m_c$. We have not computed this matrix element, since it is not relevant for the phenomenology of B decays. However, it is straightforward to compute this matrix element at zero recoil, and show that the constraint is $\chi_1(1) - 2\chi_3(1) = 0$, so that

$$\chi_1(1) = \chi_3(1) = 0. \tag{4.61}$$

Using these relations, one can derive Luke's result for the absence of $1/m_Q$ corrections to the meson matrix elements of the weak currents at zero recoil. The $\bar{B} \to D$ matrix element of the vector current at zero recoil is proportional to $h_+(1)$, and the $\bar{B} \to D^*$ matrix element of the axial current at zero recoil is proportional to $h_{A_1}(1)$. It is easy to see that $\delta h_+(1) = \delta h_{A_1}(1) = 0$ using the results derived above.

The absence of $1/m_Q$ corrections to the matrix elements of the weak currents at zero recoil allows for a precise determination of $|V_{cb}|$ from experimental semileptonic B decay data. Extrapolation of the experimental value for $\mathrm{d}\Gamma(\bar{B} \to D^* e\bar{\nu}_e)/\mathrm{d}w$ toward $w = 1$ gives

$$|V_{cb}||\mathcal{F}_{D^*}(1)| = (35.2 \pm 1.4) \times 10^{-3}, \tag{4.62}$$

where $\mathcal{F}_{D^*}(w)$ was defined in Eq. (2.87). At zero recoil, the expression for $\mathcal{F}_{D^*}(w)$ simplifies, giving $\mathcal{F}_{D^*}(1) = h_{A_1}(1)$. In the $m_Q \to \infty$ limit $\mathcal{F}_{D^*}(1) = 1$; however there are perturbative and nonperturbative corrections,

$$\mathcal{F}_{D^*}(1) = \eta_A + 0 + \delta_{1/m^2} + \cdots, \tag{4.63}$$

where η_A is the matching coefficient for the axial current, which was determined in Chapter 3 at order α_s. It has been computed to order α_s^2, and is numerically $\eta_A \simeq 0.96$. The zero in Eq. (4.63) indicates the absence of order $1/m_{c,b}$ nonperturbative corrections, and $\delta_{1/m^2} + \cdots$ stands for the nonperturbative corrections of the order of $1/m_Q^2$ and higher. Estimates of these corrections using phenomenological models like the constituent quark model lead to the expectation $\delta_{1/m^2} + \cdots \simeq -0.05$. Putting these results together, and assigning a 100% uncertainty to the model-dependent estimate of the nonperturbative effects yields the theoretical prediction

$$\mathcal{F}_{D^*}(1) = 0.91 \pm 0.05. \tag{4.64}$$

Combining this with the experimental value in Eq. (4.62) yields

$$|V_{cb}| = [38.6 \pm 1.5(\text{exp}) \pm 2.0(\text{th})] \times 10^{-3}, \tag{4.65}$$

for the $b \to c$ element of the CKM matrix.

The theoretical error in Eq. (4.64) is somewhat ad hoc. To have complete confidence that the theoretical uncertainty in the value of V_{cb} is indeed only 5%, and to try and reduce it further, it is necessary to have another high precision determination of $|V_{cb}|$ using a different method. Fortunately, as we shall see in Chapter 6, $|V_{cb}|$ can also be determined using inclusive B decays.

The zero-recoil $\bar{B} \to D$ vector current matrix element also has no order Λ_{QCD}/m_Q corrections, i.e., $h_+(1) = 1 + \mathcal{O}(\Lambda_{\text{QCD}}^2/m_Q^2)$. However, $\bar{B} \to D e \bar{\nu}_e$ is not as useful as $\bar{B} \to D^* e \bar{\nu}_e$ for determining V_{cb}. There are two reasons for this. First, the differential decay rate for $\bar{B} \to D e \bar{\nu}_e$ vanishes faster as $w \to 1$ than the differential decay rate for $\bar{B} \to D^* e \bar{\nu}_e$. This makes the extrapolation to zero recoil more difficult. Second, $\mathcal{F}_D(1)$ depends on both $h_+(1)$ and $h_-(1)$, and $h_-(1)$ does receive $\mathcal{O}(\Lambda_{\text{QCD}}/m_Q)$ corrections.

4.6 Renormalons

Suppose QCD perturbation theory is used to express some quantity f as a power series in α_s:

$$f(\alpha_s) = f(0) + \sum_{n=0}^{\infty} f_n \alpha_s^{n+1}. \tag{4.66}$$

Typically, this perturbation series for f is an asymptotic series and has zero radius of convergence. The convergence can be improved by defining the Borel transform of f,

$$B[f](t) = f(0)\delta(t) + \sum_{n=0}^{\infty} \frac{f_n}{n!} t^n, \tag{4.67}$$

which is more convergent than the original expansion in Eq. (4.66). The original series for $f(\alpha_s)$ can be recovered from the Borel transform $B[f](t)$ by the inverse Borel transform

$$f(\alpha_s) = \int_0^\infty dt\, e^{-t/\alpha_s} B[f](t). \tag{4.68}$$

If the integral in Eq. (4.68) exists, the perturbation series in Eq. (4.66) for $f(\alpha_s)$ is Borel summable, and Eq. (4.68) gives a definition for the sum of the series. While this provides a definition for the sum of the series in Eq. (4.66), it does not mean that it gives the complete, nonperturbative value for f. For example, $\exp(-1/\alpha_s)$ has the power series expansion

$$\exp(-1/\alpha_s) = 0 + 0\alpha_s + 0\alpha_s^2 + \cdots \tag{4.69}$$

whose sum is zero. If there are singularities in $B[f](t)$ along the path of integration, the Borel sum of f is ambiguous. The inverse Borel transform must be defined by deforming the contour of integration away from the singularity, and the inverse Borel transform in general depends on the deformation used.

Singularities in the Borel transform $B[f](t)$ arise from factorial growth in the coefficients f_n at high orders in perturbation theory. For example, suppose that for large n, f_n is of the order of

$$f_n \sim a w^n (n+k)! \tag{4.70}$$

The Borel transform then has a pole of order $k+1$ at $t = 1/w$:

$$B[f](t) \sim \frac{a k!}{(1 - wt)^{k+1}} + \text{less singular}. \tag{4.71}$$

One source of singularities in $B[f]$ in QCD is infrared renormalons. Infrared renormalons are ambiguities in perturbation theory arising from the fact that the gluon coupling gets strong for soft gluons. The infrared renormalons produce a factorial growth in the coefficients f_n, which gives rise to poles in the Borel transform $B[f]$. The renormalon ambiguities have a power law dependence on the momentum transfer Q^2. For example, a simple pole at $t = t_0$ in $B[f]$ introduces an ambiguity in f, depending on whether the integration contour is deformed to pass above or below the renormalon pole. The difference between

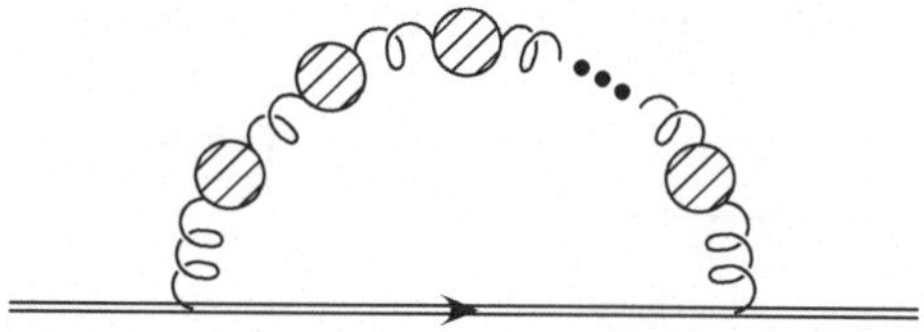

Fig. 4.1. The bubble chain sum. The blob is the gluon vacuum polarization at one loop.

the two choices is proportional to

$$\delta f \sim \oint_C dt\, e^{-t/\alpha_s(Q)} B[f](t) \sim \left(\frac{\Lambda_{\rm QCD}}{Q}\right)^{2\beta_0 t_0}, \tag{4.72}$$

where β_0 defined in Eq. (1.90) is proportional to the leading term in the QCD β function that governs the high-energy behavior of the QCD coupling constant, and the contour C encloses t_0. It is useful to write the Borel transform $B[f](t)$ in terms of the variable $u = \beta_0 t$. The form of the renormalon singularity in Eq. (4.72) then implies that a renormalon at u_0 produces an ambiguity in f that is of the order of $(\Lambda_{\rm QCD}/Q)^{2u_0}$. This ambiguity is canceled by a corresponding ambiguity in nonperturbative effects such as in the matrix elements of higher dimension operators.

Clearly, one is not able to sum the entire QCD perturbation series to determine the renormalon singularities. Typically, one sums bubble chains of the form given in Fig. 4.1. One can consider a formal limit in which the bubble chain sum is the leading term. Take QCD with N_f flavors in the limit $N_f \to \infty$, with $a = N_f \alpha_s$ held fixed. Feynman diagrams are computed to leading order in α_s, but to all orders in a. Terms in the bubble sum of Fig. 4.1 with any number of bubbles are equally important in this limit, since each additional fermion loop contributes a factor $\alpha_s N_f$, which is not small. QCD is not an asymptotically free theory in the $N_f \to \infty$ limit, so the procedure used is to write the Borel transform as a function of u but still study renormalons for positive u. The singularities in u are taken to be the renormalons for asymptotically free QCD. This procedure is a formal way of doing the bubble chain sum while neglecting other diagrams.

The Borel transform of the sum of Feynman graphs containing a single bubble chain can be readily obtained by performing the Borel transform before doing the final loop integral. In the Landau gauge, the bubble chain sum is

$$G(\alpha_s, k) = \sum_{n=0}^{\infty} \frac{i}{k^2}\left(\frac{k_\mu k_\nu}{k^2} - g_{\mu\nu}\right)(-\beta_0 \alpha_s N_f)^n [\ln(-k^2/\mu^2) + C]^n, \tag{4.73}$$

where k is the momentum flowing through the gauge boson propagator, C is a constant that depends on the particular subtraction scheme, and $\beta_0 = -1/6\pi$ is the contribution of a single fermion to the β function. In the $\overline{\rm MS}$ scheme,

$C = -5/3$. The Borel transform of Eq. (4.73) with respect to $\alpha_s N_f$ is

$$\begin{aligned} B[G](u, k) &= \frac{1}{\alpha_s N_f} \sum_{n=0}^{\infty} \frac{i}{k^2} \left(\frac{k_\mu k_\nu}{k^2} - g_{\mu\nu} \right) \frac{(-u)^n}{n!} [\ln(-k^2/\mu^2) + C]^n \\ &= \frac{1}{\alpha_s N_f} \frac{i}{k^2} \left(\frac{k_\mu k_\nu}{k^2} - g_{\mu\nu} \right) \exp[-u \ln(-k^2 e^C/\mu^2)] \\ &= \frac{1}{\alpha_s N_f} \left(\frac{\mu^2}{e^C} \right)^u \frac{i}{(-k^2)^{2+u}} (k_\mu k_\nu - k^2 g_{\mu\nu}). \end{aligned} \tag{4.74}$$

The $1/\alpha_s$ has been factored out before Borel transforming, because it will be canceled by the factor of g^2 from the gluon couplings to the external fermion line. The Borel transformed loop graphs can be computed by using the propagator in Eq. (4.74) instead of the usual gauge boson propagator in the Landau gauge:

$$(k_\mu k_\nu - k^2 g_{\mu\nu}) \frac{i}{(k^2)^2}. \tag{4.75}$$

By construction, HQET has the same infrared physics as the full QCD theory. However, because the ultraviolet physics differs in the two theories (above the scale m_Q at which the theories are matched), the coefficients of operators in the effective theory must be modified at each order in $\alpha_s(m_Q)$ to ensure that physical predictions are the same in the two theories. Such matching corrections were considered in Chapter 3.

Since the two theories coincide in the infrared, these matching conditions depend in general only on ultraviolet physics and should be independent of any infrared physics, including infrared renormalons. However, in a mass-independent renormalization scheme such as dimensional regularization with $\overline{\text{MS}}$, such a sharp separation of scales cannot be achieved. It is easy to understand why infrared renormalons appear in matching conditions. Consider the familiar case of integrating out a W boson and matching onto a four-Fermi interaction. The matching conditions at one loop involve subtracting one-loop scattering amplitudes calculated in the full and effective theories, as indicated in Fig. 4.2, where C_0 is the lowest order coefficient of the four-Fermi operator, and C_1 is the α_s correction. For simplicity, neglect all external momenta and particle masses, and consider the region of loop integration where the gluon is soft. When $k = 0$, the two theories are identical and the graphs in the two theories are identical. This

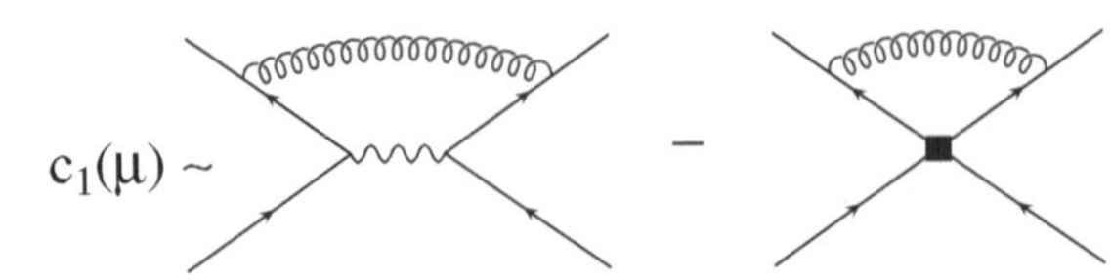

Fig. 4.2. Matching condition for the four-Fermi operator.

is the well-known statement that infrared divergences cancel in matching conditions. However, for finite (but small) k, the two theories differ at $\mathcal{O}(k^2/M_W^2)$ when one retains only the lowest dimension operators in the effective theory. Therefore, the matching conditions are sensitive to soft gluons at this order, and it is not surprising that the resulting perturbation series is not Borel summable and has renormalon ambiguities starting at $\mathcal{O}(\Lambda_{\rm QCD}^2/M_W^2)$.

However, this ambiguity is completely spurious and does not mean that the effective field theory is not well defined. Since the theory has only been defined to a fixed order, an ambiguity at higher order in $1/M_W$ is irrelevant. The renormalon ambiguity corresponded to the fact that the two theories differed in the infrared at $\mathcal{O}(k^2/M_W^2)$. When operators suppressed by an additional power of $1/M_W^2$ in the effective theory are consistently taken into account, the two theories will coincide in the infrared up to $\mathcal{O}(k^4/M_W^4)$, and any ambiguity is then pushed up to $\mathcal{O}(\Lambda_{\rm QCD}^4/M_W^4)$. Consistently including $1/M_W^4$ suppressed operators pushes the renormalon to $\mathcal{O}(\Lambda_{\rm QCD}^6/M_W^6)$, and so on. In general, a renormalon at $u = u_0$ in the coefficient function of a dimension D operator is canceled exactly by a corresponding ambiguity in matrix elements of operators of dimension $D + 2u_0$, so that physical quantities are unambiguous. This cancellation is a generic feature of all effective field theories, and it also occurs in HQET.

The HQET Lagrangian has an expansion in inverse powers of the heavy quark mass, which can be formally written as

$$\begin{aligned}
\mathcal{L} &= \mathcal{L}_0 + \mathcal{L}_1 + \mathcal{L}_2 + \cdots + \mathcal{L}_{\rm light}, \\
\mathcal{L}_0 &= \bar{Q}_v (iD\cdot v) Q_v - \delta m \bar{Q}_v Q_v,
\end{aligned} \tag{4.76}$$

on scaling out the phase factor $\exp(-im_0 v\cdot x)$ from the heavy quark field. Here m_0 is a mass that can differ from m_Q by an amount of order $\Lambda_{\rm QCD}$, $\mathcal{L}_{\rm light}$ is the QCD Lagrangian for the light quarks and gluons, Q_v is the heavy quark field, and $\mathcal{L}_k$ are terms in the effective Lagrangian for the heavy quark that are of order $1/m_0^k$. There are two mass parameters for the heavy quark in Eq. (4.76), the expansion parameter of HQET m_0, and the residual mass term δm. The two parameters are not independent; one can make the redefinition $m_0 \to m_0 + \Delta m$, $\delta m \to \delta m - \Delta m$. A particularly convenient choice is to adjust m_0 so that the residual mass term δm vanishes. Most HQET calculations are done with this choice of m_0, and this is the choice we have used so far in this book, but it is easy to show that the same results are obtained with a different choice of m_0. The HQET mass m_0 when $\delta m = 0$ is often referred to in the literature as the pole mass m_Q, and we will follow this practice here.

Like all effective Lagrangians, the HQET Lagrangian is nonrenormalizable, so a specific regularization prescription must be included as part of the definition of the effective theory. An effective field theory is used to compute physical quantities in a systematic expansion in a small parameter, and the effective

Lagrangian is expanded in this small parameter. The expansion parameter of the HQET is Λ_{QCD}/m_0. One can then use "power counting" to determine what terms in the effective theory are relevant to a given order in the $1/m_0$ expansion. For example, to second order in $1/m_0$, one needs to study processes to first order in $\mathcal{L}_2$, and to second order in $\mathcal{L}_1$. It is useful to have a renormalization procedure that preserves the power counting. We choose to use dimensional regularization with $\overline{\text{MS}}$, and nonperturbative matrix elements must be interpreted in this scheme. A nonperturbative calculation of a matrix elements, e.g., using lattice Monte Carlo methods, can be converted to $\overline{\text{MS}}$ by means of a perturbative matching procedure.

There is a renormalon in the relation between the renormalized mass at short distances (such as the $\overline{\text{MS}}$ mass $\bar{m}_Q$) and the pole mass of the heavy quark at $u = 1/2$, which produces an ambiguity of the order of Λ_{QCD} in the relation between the pole mass and the $\overline{\text{MS}}$ mass. The heavy quark mass in HQET and the $\overline{\text{MS}}$ mass at short distances are parameters in the Lagrangian that must be determined from experiment. Any scheme can be used to compute physical processes, though one scheme might be more advantageous for a particular computation. The $\overline{\text{MS}}$ mass at short distances is useful in computing high-energy processes. However, there is no advantage to using the "short distance" mass (such as the running $\overline{\text{MS}}$ mass) in HQET. In fact, from the point of view of HQET, this is inconvenient. The effective Lagrangian in Eq. (4.76) is an expansion in inverse powers of m_0. Power counting in $1/m_0$ in the effective theory is only valid if δm is of the order of one (or smaller) in m_0, i.e., only if δm remains finite in the infinite mass limit $m_0 \to \infty$. When m_0 is chosen to be the $\overline{\text{MS}}$ mass the residual mass term δm is of the order of m_0 (up to logarithms). This spoils the $1/m_0$ power counting of HQET, mixes the α_s and $1/m_0$ expansions, and breaks the heavy flavor symmetry. For example, using m_0 to be the $\overline{\text{MS}}$ mass at $\mu = m_0$, one finds at one loop that

$$\delta m = \frac{4}{3\pi}\alpha_s m_0. \tag{4.77}$$

In $b \to c$ decays, including this residual mass term in the heavy c-quark Lagrangian causes $1/m_c$ operators such as $\bar{c}_{v'}\overleftarrow{\not{D}}\Gamma b_v/m_c$ to produce effects that are suppressed by α_s rather than Λ_{QCD}/m_c. While physical quantities calculated in this way must be the same as those calculated by using the pole mass, it unnecessarily complicates the power counting to use a definition for m_0 that leaves a residual mass term that is not finite in the $m_0 \to \infty$ limit. Better choices for the expansion parameter of HQET are the heavy meson mass (with δm of the order of Λ_{QCD}), and the pole mass (with $\delta m = 0$).

The $\overline{\text{MS}}$ mass at short distances can be determined (in principle) from experiment without any renormalon ambiguities proportional to Λ_{QCD}. The $\overline{\text{MS}}$ quark mass can be related to other definitions of the quark mass by using QCD perturbation theory. The connection between the Borel-transformed pole mass and the $\overline{\text{MS}}$ mass is

$$B[m_Q](u) = \bar{m}_Q\delta(u) + \frac{\bar{m}_Q}{3\pi N_f}\left[\left(\frac{\mu^2}{\bar{m}_Q^2}\right)^u e^{-uC}6(1-u)\frac{\Gamma(u)\Gamma(1-2u)}{\Gamma(3-u)} - \frac{3}{u} + R_{\Sigma_1}(u)\right], \tag{4.78}$$

where $\bar{m}_Q$ is the renormalized $\overline{\text{MS}}$ mass at the subtraction point μ, and the constant $C = -5/3$ and the function $R_{\Sigma_1}(u)$ have no singularities at $u = 1/2$. Equation (4.78) has a renormalon singularity at $u = 1/2$, which is the leading infrared renormalon in the pole mass. Writing $u = 1/2 + \Delta u$, we have

$$B[m_Q](u = 1/2 + \Delta u) = -\frac{2\mu e^{-C/2}}{3\pi N_f\, \Delta u} + \cdots, \tag{4.79}$$

where the ellipses denote terms regular at $\Delta u = 0$. We will only work to leading order in $1/m_O$, so poles to the right of $u = 1/2$, which are related to ambiguities at higher order in $1/m_O$, are irrelevant. Although m_Q is formally ambiguous at Λ_{QCD}, we have argued that physical quantities that depend on m_Q are unambiguously predicted in HQET. We now demonstrate this explicitly for a ratio of form factors in Λ_b semileptonic decay.

The matrix element of the vector current for the semileptonic decay $\Lambda_b \to \Lambda_c e\bar{\nu}_e$ decay is parameterized by the three decay form factors $f_{1-3}(w)$ defined in Eq. (4.27). In the limit $m_b,\ m_c \to \infty$, and at lowest order in α_s, the form factors f_2 and f_3 vanish. We will consider α_s and $1/m_c$ corrections, but work in the $m_b \to \infty$ limit. Consider the ratio $r_f = f_2/f_1$, which vanishes at lowest order in α_s and $1/m_c$. The corrections to r_f can be written in the form

$$r_f(\alpha_s, w) \equiv \frac{f_2(w)}{f_1(w)} = -\frac{\bar{\Lambda}_\Lambda}{m_c}\frac{1}{(1+w)} + f_r(\alpha_s, w), \tag{4.80}$$

where the function $f_r(\alpha_s, w)$ is a perturbatively calculable matching condition from the theory above $\mu = m_c$ to the effective theory below $\mu = m_c$, and the $\bar{\Lambda}_\Lambda$ term arises from $1/m_c$ suppressed operators in HQET. At one loop (see Problem 5 of Chapter 3),

$$f_r(\alpha_s, w) = -\frac{2\alpha_s}{3\pi}\frac{1}{\sqrt{w^2-1}}\ln\left(w + \sqrt{w^2-1}\right). \tag{4.81}$$

The ratio $r_f = f_2/f_1$ is an experimentally measurable quantity and does not have a renormalon ambiguity. The standard form for r_f in Eq. (4.80) is obtained by using HQET with the pole mass as the expansion parameter. The HQET parameter $\bar{\Lambda}_\Lambda$ is the baryon mass in the effective theory, i.e., it is the baryon mass m_{Λ_c} minus the pole mass of the c quark. The pole mass has the leading renormalon

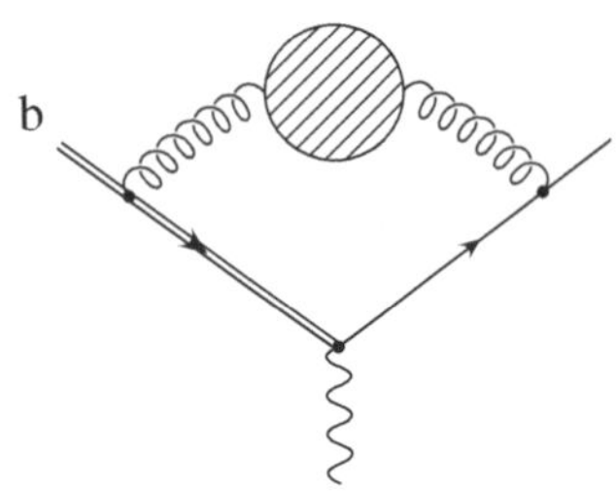

Fig. 4.3. The bubble chain sum for the radiative correction to the vector current form factors.

ambiguity at $u = 1/2$ given in Eq. (4.79), which produces an ambiguity in the $1/m_c$ contribution to f_2/f_1 given by the first term in Eq. (4.80). There must therefore also be a renormalon at $u = 1/2$ in the radiative correction to f_2/f_1 given by the second term in Eq. (4.80). It is straightforward to show that this is indeed the case.

The Borel-transformed series $B[f_r](u, w)$ in the $1/N_f$ expansion is easily calculated from the graph in Fig. 4.3, using the Borel-transformed propagator in Eq. (4.74). The Borel transform of the Feynman diagram is

$$B[\text{graph}] = \frac{i}{\alpha_s N_f}\frac{4}{3}g^2\left(\frac{\mu^2}{e^C}\right)^u \times \int \frac{d^4k}{(2\pi)^4}\frac{\gamma^\nu(m_c\not{v}' + \not{k} + m_c)\gamma^\alpha v^\mu(k_\mu k_\nu - k^2 g_{\mu\nu})}{(k^2 + 2m_c k\cdot v')(-k^2)^{2+u}k\cdot v}. \tag{4.82}$$

The radiative correction to f_2 (which determines f_r) is obtained from the terms in Eq. (4.82) that are proportional to v^α. Combining denominators using Eq. (1.45) and Eq. (3.6), extracting the terms proportional to v^α and performing the momentum integral, we obtain

$$B[f_r](u, w) = \frac{4(u-2)}{3\pi N_f(1+u)}\left(\frac{\mu^2}{e^C}\right)^u m_c \times \int_0^\infty d\lambda \int_0^1 dx \frac{(1-x)^{1+u}\,x}{\left[\lambda^2 + 2\lambda m_c x w + m_c^2 x^2\right]^{1+u}}. \tag{4.83}$$

Rescaling $\lambda \to x m_c \lambda$ and performing the x integral gives

$$B[f_r](u, w) = \frac{4}{3\pi N_f}\left(\frac{\mu^2}{m_c^2 e^C}\right)^u \frac{(u-2)\Gamma(1-2u)\Gamma(1+u)}{\Gamma(3-u)} \times \int_0^\infty d\lambda \frac{1}{[\lambda^2 + 2\lambda w + 1]^{1+u}}. \tag{4.84}$$

This expression has a pole at $u = 1/2$. Expanding in $\Delta u = u - 1/2$ gives

$$\begin{aligned} B[f_r](u = 1/2 + \Delta u, w) \\ &= \frac{2\mu}{3\pi N_f m_c e^{C/2}} \frac{1}{\Delta u} \int_0^\infty d\lambda \frac{1}{[\lambda^2 + 2\lambda w + 1]^{3/2}} + \cdots \\ &= \frac{2\mu}{3\pi N_f m_c e^{C/2}} \frac{1}{\Delta u} \frac{1}{1+w}, \end{aligned} \tag{4.85}$$

where the ellipsis denotes terms that are regular at $u = 1/2$.

The Borel singularity in Eq. (4.85) cancels the singularity in the first term of Eq. (4.80) at all values of w, so that the ratio of form factors $r_f(\alpha_s, w) = f_2(w)/f_1(w)$ has no renormalon ambiguities. Therefore the standard HQET computation of the $1/m_c$ correction to f_2/f_1 using the pole mass and the standard definition of $\bar{\Lambda}_\Lambda$ gives an unambiguous physical prediction for the ratio of form factors.

The cancellation of renormalon ambiguities has been demonstrated by explicit computation in this example, but the result holds in general.

4.7 $v \cdot A = 0$ gauge

Calculations in HQET can be performed in almost any gauge. However, in the $v \cdot A = 0$ gauge, HQET perturbation theory is singular. Consider tree-level Qq elastic scattering in the rest frame $v = v_r$. In HQET, an on-shell heavy quark has a four velocity v and a residual momentum k that satisfies $v \cdot k = 0$. Suppose the initial heavy quark has zero residual momentum and the final quark has residual momentum $k = (0, \mathbf{k})$. The tree-level Feynman diagram in Fig. 4.4 gives the Qq scattering amplitude

$$\mathcal{M} = -g^2 \bar{u}_Q T^A u_Q \frac{i}{\mathbf{k}^2} \bar{u}_q T^A \not{v} u_q, \tag{4.86}$$

in the Feynman or Landau gauge, where u_Q and u_q are the heavy and light quark spinors, respectively. The current conservation equation $\bar{u}_q \not{k} u_q = 0$ was used to simplify the result.

In the $v \cdot A = 0$ gauge, the gluon propagator is

$$\frac{-i}{k^2 + i\varepsilon} \left[g_{\mu\nu} - \frac{1}{v \cdot k}(k_\mu v_\nu + v_\mu k_\nu) + \frac{1}{(v \cdot k)^2} k_\mu k_\nu \right]. \tag{4.87}$$

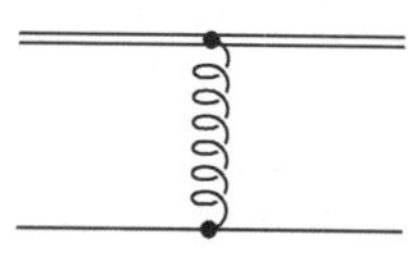

Fig. 4.4. Heavy quark + light quark scattering amplitude at tree level.

The heavy quark kinetic energy cannot be treated as a perturbation in this gauge, because then $v \cdot k = 0$ and the gluon propagator is ill defined. Including the heavy quark kinetic energy in the Lagrangian, the residual momentum of the outgoing heavy quark becomes $k^\mu = (\mathbf{k}^2/2m_Q, \mathbf{k})$ and $v \cdot k = \mathbf{k}^2/2m_Q$ is not zero. Note that the factors of $1/(v \cdot k)$ in Eq. (4.87) lead to $2m_Q/\mathbf{k}^2$ terms in the gluon propagator, so that the $v \cdot A = 0$ gauge can mix different orders in the $1/m_Q$ expansion.

It is instructive to see how the scattering amplitude in Eq. (4.86) arises in the $v \cdot A = 0$ gauge. The amplitude comes from the QQA vertex that is due to the heavy quark kinetic energy term $-\bar{Q}_v D_\perp^2/(2m_Q)Q_v$. Although this is a $1/m_Q$ term in the Lagrangian, it can contribute to a leading-order amplitude in the $v \cdot A = 0$ gauge. The Feynman rule for $Q_v(k') \to Q_v(k) + A_\mu$ vertex arising from an insertion of the kinetic energy operator is $i(g/2m_Q)(k_\perp + k'_\perp)_\mu = i(g/2m_Q)(k+k')_\mu - i(g/2m_Q)v \cdot (k+k')v_\mu$. In the case we are considering, v is chosen so that $k' = 0$. The part proportional to v_μ doesn't contribute, since $v \cdot A = 0$. Since $\bar{u}_q \not{k} u_q = 0$ only the $v_\mu k_\nu + v_\nu k_\mu$ term in the gluon propagator contributes, and one can show that it reproduces Eq. (4.86) for large values of m_Q.

In the $v \cdot A = 0$ gauge the heavy quark kinetic energy must be considered as a leading operator for on-shell scattering processes, because we have just seen that it is the QQA vertex from this $1/m_Q$ operator that gives rise to the leading Qq on-shell scattering amplitude.

4.8 NRQCD

HQET is not the appropriate effective field theory for systems with more than one heavy quark. In HQET the heavy quark kinetic energy is neglected. It occurs as a small $1/m_Q$ correction. At short distances the static potential between heavy quarks is determined by one gluon exchange and is a Coulomb potential. For a $Q\bar{Q}$ pair in a color singlet, it is an attractive potential, and the heavy quark kinetic energy is needed to stabilize a $Q\bar{Q}$ meson. For $Q\bar{Q}$ hadrons (i.e., quarkonia) the kinetic energy plays a very important role, and it cannot be treated as a perturbation.

In fact the problem is more general than this. Consider, for example, trying to calculate low-energy QQ scattering in the center of a mass frame using HQET. Setting $v = v_r$ for each heavy quark, and using initial and final residual momenta $k_\pm = (0, \pm\mathbf{k})$ and $k'_\pm = (0, \pm\mathbf{k}')$ respectively, we find the one-loop Feynman diagram, Fig. 4.5, gives rise to a loop integral,

$$\int \frac{d^n q}{(2\pi)^n} \frac{i}{(q^0 + i\varepsilon)} \frac{i}{(-q^0 + i\varepsilon)} \frac{i}{(q + k_+)^2 + i\varepsilon} \frac{i}{(q + k'_+)^2 + i\varepsilon}. \tag{4.88}$$

The q^0 integration is ill defined because it has poles above and below the real axis

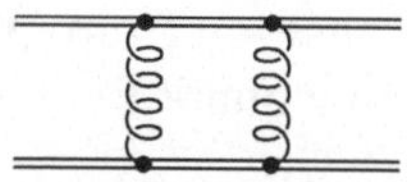

Fig. 4.5. One-loop contribution to QQ scattering.

at $q^0 = \pm i\varepsilon$. This problem is cured by not treating the heavy quark kinetic energy as a perturbation but including it in the leading-order terms. Then the denominators of the two heavy quark propagators become $E + q^0 - \mathbf{q}^2/2m_Q + i\varepsilon$ and $E - q^0 - \mathbf{q}^2/2m_Q + i\varepsilon$, where $E = \mathbf{k}^2/2m_Q = \mathbf{k}'^2/2m_Q$. Closing the q^0 contour in the upper half-plane, we find Eq. (4.88) is dominated (for large m_Q) by the residue of the pole at $q^0 = E - \mathbf{q}^2/2m_Q + i\varepsilon$ and is proportional to m_Q. That is why we obtained an infinite answer for Eq. (4.88) by using the $m_Q \to \infty$ limit of the fermion propagators.

Properties of quarkonia are usually predicted as a power series in v/c, where v is the magnitude of the relative $Q\bar{Q}$ velocity and c is the speed of light. For these systems the appropriate limit of QCD to examine is the $c \to \infty$ limit. In this limit the QCD Lagrangian becomes an effective field theory called NRQCD. For finite c there are corrections suppressed by powers of $1/c$. In particle physics we usually set $\hbar = c = 1$. Making the factors of c explicit, we find the QCD Lagrangian density is

$$\mathcal{L}_{\text{QCD}} = -\frac{1}{4} G^B_{\mu\nu} G^{B\mu\nu} - c\bar{Q}(i\not{D} - m_Q c)Q. \tag{4.89}$$

In the above the zero component of a partial derivative is

$$\partial_0 = \frac{1}{c}\frac{\partial}{\partial t}, \tag{4.90}$$

and D is the covariant derivative

$$D_\mu = \partial_\mu + \frac{ig}{c} A^B_\mu T^B. \tag{4.91}$$

The gluon field strength tensor $G^B_{\mu\nu}$ is defined in the usual way except that $g \to g/c$.

Although c is explicit, $\hbar$ has been set to unity. All dimensionful quantities can be expressed in units of length $[x]$ and time $[t]$, i.e., $[E] \sim 1/[t]$ and $[p] \sim 1/[x]$. The Lagrangian $L = \int d^3x \mathcal{L}$ has units of $1/[t]$ since the action $S = \int \mathcal{L} dt$ is dimensionless. It is straightforward to deduce that the gluon field has units $[A] \sim 1/\sqrt{[x][t]}$ and the strong coupling $g \sim \sqrt{[x]/[t]}$. The fermion field has units $[\psi] \sim 1/[x]^{3/2}$ while its mass has units $[m_Q] \sim [t]/[x]^2$. With these units $m_Q c^2$ has dimensions of energy and the strong fine structure constant $\alpha_s = g^2/4\pi c$ is dimensionless.

For the fermion field Q the transition from QCD to NRQCD is analogous to the derivation of HQET. The heavy quark field is rewritten as

$$Q = e^{-im_Q c^2 t}\left[1 + \frac{i\not{D}_\perp}{m_Q c} + \cdots\right]\begin{pmatrix}\psi \\ 0\end{pmatrix}, \tag{4.92}$$

where ψ is a two-component Pauli spinor. Using this field redefinition, we find the part of the QCD Lagrange density involving Q becomes

$$\mathcal{L}_\psi = \psi^\dagger\left[i\left(\frac{\partial}{\partial t} + igA_0^B T^B\right) + \frac{\boldsymbol{\nabla}^2}{2m_Q}\right]\psi + \cdots, \tag{4.93}$$

where the ellipses denote terms suppressed by powers of $1/c$. Note that the heavy quark kinetic energy is now leading order in $1/c$. The replacement $g \to g/c$ was necessary to have a sensible $c \to \infty$ limit.

Among the terms suppressed by a single power of $1/c$ is the gauge completion of the kinetic energy:

$$\mathcal{L}_{\text{int}} = \frac{ig}{2m_Q c}\mathbf{A}^C[\psi^\dagger T^C \boldsymbol{\nabla}\psi - (\boldsymbol{\nabla}\psi)^\dagger T^C \psi]. \tag{4.94}$$

There is also a $1/c$ term involving the color magnetic field $\mathbf{B}^C = \boldsymbol{\nabla} \times \mathbf{A}^C$.

It is convenient to work in Coulomb gauge, $\boldsymbol{\nabla} \cdot \mathbf{A}^C = 0$. Then the part of the action that involves the gluon field strength tensor and is quadratic in the gauge fields simplifies to

$$\begin{aligned} -\frac{1}{4}\int d^3x G^C_{\mu\nu}G^{C\mu\nu} &\to \frac{1}{2}\int d^3x G^C_{0i}G^C_{0i} - \frac{1}{4}\int d^3x G^C_{ij}G^C_{ij} \\ &= \frac{1}{2}\int d^3x (\partial_i A_0^C)^2 + (\partial_0 A_i^C)^2 - (\partial_i A_j^C)^2 + \text{ non-Abelian terms.} \end{aligned} \tag{4.95}$$

The non-Abelian terms are suppressed by factors of $1/c$. [The above derivation implicitly assumes that $m_Q v^2 \gg \Lambda_{\text{QCD}}$.]

In Eq. (4.95), the zero component of the gauge field has no time derivatives. Therefore, it does not represent a propagating degree of freedom. Neglecting terms suppressed by factors of $1/c$, the Lagrangian only contains terms quadratic and linear in the field A_0^C. Hence the functional integral over A_0^C can be performed exactly by completing the square. The effects of A_0^C exchange are then reproduced by an instantaneous potential $V(\mathbf{x}, \mathbf{y})$ that is proportional to the Fourier transform of the momentum–space propagator,

$$V(\mathbf{x}, \mathbf{y}) = g^2 \int \frac{d^3k}{(2\pi)^3} e^{i\mathbf{k}\cdot(\mathbf{x}-\mathbf{y})}\frac{1}{\mathbf{k}^2} = \frac{g^2}{4\pi|\mathbf{x} - \mathbf{y}|}. \tag{4.96}$$

The transverse gluons $\mathbf{A}^C$ do not couple to the quarks at leading order in the $1/c$ expansion. Neglecting terms suppressed by $1/c$, we find the effective Lagrangian for the interaction of nonrelativistic quarks is

$$L_{\text{NRQCD}} = \int d^3x\, \psi^\dagger \left(i\frac{\partial}{\partial t} + \frac{\boldsymbol{\nabla}^2}{2m_Q} \right) \psi - \int d^3x_1 \int d^3x_2 \psi^\dagger(\mathbf{x}_1, t) T^A \psi(\mathbf{x}_1, t) V(\mathbf{x}_1, \mathbf{x}_2) \psi^\dagger(\mathbf{x}_2, t) T^A \psi(\mathbf{x}_2, t). \tag{4.97}$$

The Hamiltonian

$$H = \int d^3x \psi^\dagger i \frac{\partial}{\partial t} \psi - L \tag{4.98}$$

has the familiar form used in nonrelativistic many-body theory. When restricting one's attention to the two heavy quark sector, the effective theory reduces to ordinary nonrelativistic quantum mechanics.

4.9 Problems

1. For any doublet of heavy hadrons $H_\pm^{(Q)}$ with spins $j_\pm = s_\ell \pm 1/2$, show that

$$m_{H_\pm^{(Q)}} = m_Q + \bar{\Lambda}_H - \frac{\lambda_{H,1}}{2m_Q} \pm n_\mp \frac{\lambda_{H,2}}{2m_Q},$$

where $n_\pm = 2j_\pm + 1$ and $\lambda_{H,1}$ and $\lambda_{H,2}$ are defined in Eqs. (4.23). We have inserted an extra subscript H because the values of the matrix elements depend on the particular doublet.

2. For the ground-state doublet of mesons, let $\{\bar{\Lambda}_H, \lambda_{H,1}, \lambda_{H,2}\} = \{\bar{\Lambda}, \lambda_1, \lambda_2\}$ and for the excited $s_\ell = 3/2$ mesons let $\{\bar{\Lambda}_H, \lambda_{H,1}, \lambda_{H,2}\} = \{\bar{\Lambda}^*, \lambda_1^*, \lambda_2^*\}$. Show that

$$\bar{\Lambda}^* - \bar{\Lambda} = \frac{m_b(\bar{m}_B^* - \bar{m}_B) - m_c(\bar{m}_D^* - \bar{m}_D)}{m_b - m_c},$$

$$\lambda_1^* - \lambda_1 = 2m_c m_b \frac{(\bar{m}_B^* - \bar{m}_B) - (\bar{m}_D^* - \bar{m}_D)}{m_b - m_c},$$

where

$$\bar{m}_H = \frac{n_- m_{H_-} + n_+ m_{H_+}}{n_+ + n_-}.$$

3. In Problems 6–9 of Chapter 2, the leading $m_Q \to \infty$ predictions for the $\bar{B} \to D_1 e \bar{\nu}_e$ and $\bar{B} \to D_2^* e \bar{\nu}_e$ form factors were derived. In this problem, the $1/m_Q$ corrections are included.

(a) For $\bar{B} \to D_1$ and $\bar{B} \to D_2^*$ matrix elements, argue that

$$\bar{c}_{v'}\, i\overleftarrow{D}_\lambda \Gamma\, b_v = \text{Tr}\{S_{\sigma\lambda}^{(c)} \bar{F}_{v'}^\sigma \Gamma H_v^{(b)}\},$$

$$\bar{c}_{v'}\, \Gamma i D_\lambda\, b_v = \text{Tr}\{S_{\sigma\lambda}^{(b)} \bar{F}_{v'}^\sigma \Gamma H_v^{(b)}\},$$

where

$$S_{\sigma\lambda}^{(Q)} = v_\sigma \left[\tau_1^{(Q)} v_\lambda + \tau_2^{(Q)} v'_\lambda + \tau_3^{(Q)} \gamma_\lambda \right] + \tau_4^{(Q)} g_{\sigma\lambda},$$

and the functions $\tau_i^{(Q)}$ depend on w. (They are not all independent.)

(b) Show that the heavy quark equation of motion implies

$$w\tau_1^{(c)} + \tau_2^{(c)} - \tau_3^{(c)} = 0,$$
$$\tau_1^{(b)} + w\tau_2^{(b)} - \tau_3^{(b)} + \tau_4^{(b)} = 0.$$

(c) Further relations between the τ's follow from

$$i\partial_\nu(\bar{c}_{v'}\,\Gamma\, b_v) = (\bar{\Lambda} v_\nu - \bar{\Lambda}^* v'_\nu)\bar{c}_{v'}\,\Gamma\, b_v.$$

Show that this equation implies the relations

$$\begin{aligned}
\tau_1^{(c)} + \tau_1^{(b)} &= \bar{\Lambda}\tau,\\
\tau_2^{(c)} + \tau_2^{(b)} &= -\bar{\Lambda}^*\tau,\\
\tau_3^{(c)} + \tau_3^{(b)} &= 0,\\
\tau_4^{(c)} + \tau_4^{(b)} &= 0,
\end{aligned}$$

where τ was defined in Problem 9 of Chapter 2. The relations in parts (b) and (c) imply that all the $\tau_j^{(Q)}$'s can be expressed in terms of $\tau_1^{(c)}$ and $\tau_2^{(c)}$.

(d) Using the results from parts (a)–(c), show that the corrections to the currents give the following corrections to the form factors:

$$\begin{aligned}
\sqrt{6}\,\delta f_A &= -\epsilon_b(w-1)[(\bar{\Lambda}^* + \bar{\Lambda})\tau - (2w+1)\tau_1 - \tau_2]\\
&\quad - \epsilon_c[4(w\bar{\Lambda}^* - \bar{\Lambda})\tau - 3(w-1)(\tau_1 - \tau_2)],\\
\sqrt{6}\,\delta f_{V_1} &= -\epsilon_b(w^2-1)[(\bar{\Lambda}^* + \bar{\Lambda})\tau - (2w+1)\tau_1 - \tau_2]\\
&\quad - \epsilon_c[4(w+1)(w\bar{\Lambda}^* - \bar{\Lambda})\tau - 3(w^2-1)(\tau_1 - \tau_2)],\\
\sqrt{6}\,\delta f_{V_2} &= -3\epsilon_b[(\bar{\Lambda}^* + \bar{\Lambda})\tau - (2w+1)\tau_1 - \tau_2] - \epsilon_c[(4w-1)\tau_1 + 5\tau_2)],\\
\sqrt{6}\,\delta f_{V_3} &= \epsilon_b(w+2)[(\bar{\Lambda}^* + \bar{\Lambda})\tau - (2w+1)\tau_1 - \tau_2]\\
&\quad + \epsilon_c[4(w\bar{\Lambda}^* - \bar{\Lambda})\tau + (2+w)\tau_1 + (2+3w)\tau_2],
\end{aligned}$$

for $\bar{B} \to D_1 e\bar{\nu}_e$. For $\bar{B} \to D_2^* e\bar{\nu}_e$ show that the corrections to the form factors are

$$\begin{aligned}
\delta k_V &= -\epsilon_b[(\bar{\Lambda}^* + \bar{\Lambda})\tau - (2w+1)\tau_1 - \tau_2] - \epsilon_c[\tau_1 - \tau_2],\\
\delta k_{A_1} &= -\epsilon_b(w-1)[(\bar{\Lambda}^* + \bar{\Lambda})\tau - (2w+1)\tau_1 - \tau_2] - \epsilon_c(w-1)[\tau_1 - \tau_2],\\
\delta k_{A_2} &= -2\epsilon_c\tau_1,\\
\delta k_{A_3} &= \epsilon_b[(\bar{\Lambda}^* + \bar{\Lambda})\tau - (2w+1)\tau_1 - \tau_2] - \epsilon_c[\tau_1 + \tau_2].
\end{aligned}$$

Here $\epsilon_c = 1/(2m_c)$, $\epsilon_b = 1/(2m_b)$ and $\tau_1 = \tau_1^{(c)}$, $\tau_2 = \tau_2^{(c)}$.

(e) The zero-recoil matrix elements of the weak current are determined by $f_{V_1}(1)$. The $1/m_Q$ corrections to the current imply that

$$\sqrt{6} f_{V_1}(1) = -8\epsilon_c(\bar{\Lambda}^* - \bar{\Lambda})\tau(1).$$

Show that the $1/m_Q$ corrections to the states do not alter this relation.

4. Explain why the χ_1's from the charm and bottom quark kinetic energies are the same.
5. Show that the $\bar{B}^* \to D^*$ matrix element implies that $\chi_1(1) - 2\chi_3(1) = 0$ for the $1/m_Q$ corrections to the $\bar{B} \to D^{(*)}$ form factors that arise from the chromomagnetic term in the Lagrangian.
6. Verify Eq. (4.77) for the relation between the $\overline{\text{MS}}$ mass and the pole mass.
7. Calculate the order $\Lambda_{\text{QCD}}/m_{c,b}$ corrections to the form factor ratios R_1 and R_2 defined in Chapter 2. Express the result in terms of $\bar{\Lambda}$, ξ_3 and χ_{1-3}.

4.10 References

$1/m_Q$ corrections to the Lagrangian were computed in:

E. Eichten and B. Hill, Phys. Lett. B243 (1990) 427
A.F. Falk, M.E. Luke, and B. Grinstein, Nucl. Phys. B357 (1991) 185

Luke's theorem was proved for meson decays in:

M.E. Luke, Phys. Lett. B252 (1990) 447

and extended to the general case in:

C.G. Boyd and D.E. Brahm, Phys. Lett. B257 (1991) 393

$1/m_Q$ corrections to baryon form factors were discussed in:

H. Georgi, B. Grinstein, and M.B. Wise, Phys. Lett. B252 (1990) 456

Reparameterization invariance was formulated in:

M.E. Luke and A.V. Manohar, Phys. Lett. B286 (1992) 348

for an application see:

M. Neubert, Phys. Lett. B306 (1993) 357

For an early discussion on determining V_{cb} from exclusive decays, see:

M. Neubert, Phys. Lett. B264 (1991) 455

Formulating HQET with a residual mass term was discussed in:

A.F. Falk, M. Neubert, and M.E. Luke, Nucl. Phys. B388 (1992) 363

QCD renormalons were studied in:

G. 't Hooft, in *The Whys of Subnuclear Physics*, edited by A. Zichichi, New York, Plenum, 1978

The renormalon ambiguity in the pole mass was studied in:

I.I. Bigi, M.A. Shifman, N.G. Uraltsev, and A.I. Vainshtein, Phys. Rev. D50 (1994) 2234
M. Beneke and V.M. Braun, Nucl. Phys. B426 (1994) 301

Cancellation of the renormalon ambiguity between perturbative and nonperturbative corrections for HQET was shown in:

M. Beneke, V.M. Braun, and V.I. Zakharov, Phys. Rev. Lett. 73 (1994) 3058
M.E. Luke, A.V. Manohar, and M.J. Savage, Phys. Rev. D51 (1995) 4924
M. Neubert and C.T. Sachrajda, Nucl. Phys. B438 (1995) 235

For a recent review on renormalons see:

M. Beneke, hep-ph/9807443

Order $1/m_Q^2$ corrections to $\bar{B} \to D^{(*)}$ and $\Lambda_b \to \Lambda_c$ semileptonic decay form factors are considered in:

A.F. Falk and M. Neubert, Phys. Rev. D47 (1993) 2965, D47 (1993) 2982

For the HQET Lagrangian at order $1/m_Q^2$, see:

T. Mannel, Phys. Rev. D50 (1994) 428
I.I. Bigi, M.A. Shifman, N.G. Uraltsev, and A.I. Vainshtein, Phys. Rev. D52 (1995) 196
B. Blok, J.G. Korner, D. Pirjol, and J.C. Rojas, Nucl. Phys. B496 (1997) 358
C. Bauer and A.V. Manohar, Phys. Rev. D57 (1998) 337

For the HQET Lagrangian at order $1/m_Q^3$, see:

A.V. Manohar, Phys. Rev. D56 (1997) 230
A. Pineda and J. Soto, hep-ph/9802365

NRQCD was discussed in:

W.E. Caswell and G.P. Lepage, Phys. Lett. B167 (1986) 437
G.P. Lepage and B.A. Thacker, Presented at the Int. Symp. on Quantum Field Theory on the Lattice, Seillac, France 1987
G.T. Bodwin, E. Braaten, and G.P. Lepage Phys. Rev. D51 (1995) 1125
P. Labelle, Phys. Rev. D58 (1998) 093013
B. Grinstein and I.Z. Rothstein, Phys. Rev. D57 (1998) 78

$1/m_Q$ corrections to the form factors for $\bar{B}$ decay to excited charmed mesons were discussed in:

A. Leibovich, Z. Ligeti, I. Stewart, and M.B. Wise, Phys. Rev. Lett. 78 (1997) 3995, Phys. Rev. D57 (1998) 308

See also:

M. Neubert, Phys. Lett. B418 (1998) 173

5
Chiral perturbation theory

In Sec. 1.4 we discussed how to formulate an effective chiral Lagrangian for the self-interactions of low-momentum pseudo-Goldstone bosons, such as the pion. Chiral Lagrangians can also be used to describe the interactions of pions with hadrons containing a heavy quark. The use of chiral perturbation theory is valid for these interactions as long as the pion is soft, that is, has momentum $p \ll \Lambda_{CSB}$. Chiral perturbation theory for heavy hadrons makes use of spontaneously broken $SU(3)_L \times SU(3)_R$ chiral symmetry on the light quarks, and spin-flavor symmetry on the heavy quarks. In this chapter, we study the implications of the combination of chiral and heavy quark symmetries for heavy hadron pion interactions.

5.1 Heavy mesons

In this section, we will obtain the chiral Lagrangian that describes the low-momentum interactions of the π, K, and η with the ground state $s_\ell = \frac{1}{2}$ spin symmetry doublet of heavy mesons, P_a and P_a^*. Some applications of the chiral Lagrangian are described later in this chapter. The chiral Lagrangian for other heavy hadron multiplets, e.g., heavy baryons, can be obtained similarly and is left to the problems at the end of the chapter.

As was noted in Chapter 2, we can combine the P_a and P_a^* fields into a 4×4 matrix,

$$H_a = \frac{(1 + \not{v})}{2}\left[P_a^{*\mu}\gamma_\mu + i P_a \gamma_5\right], \tag{5.1}$$

that transforms under the unbroken $SU(3)_V$ subgroup of chiral symmetry as an antitriplet

$$H_a \to H_b V_{ba}^\dagger, \tag{5.2}$$

and transforms as a doublet

$$H_a \to D_Q(R) H_a, \tag{5.3}$$

under heavy quark spin symmetry. In Chapter 2, the fields P, P^* and the matrix H were also labeled by the flavor and velocity of the heavy quark. In this chapter, we are mainly concerned with the light quark dynamics, and so these labels are suppressed whenever possible.

The Lagrangian for the strong interactions of the P and P^* with low-momentum pseudo-Goldstone bosons should be the most general one consistent with the chiral and heavy quark symmetries defined in Eqs. (5.2) and (5.3), and it should contain at leading order the least number of derivatives and insertions of the light quark mass matrix. Fields such as P and P^*, which are not Goldstone bosons, are generically referred to as matter fields. Matter fields have a well-defined transformation rule under the unbroken vector $SU(3)_V$ symmetry, but they do not necessarily form representations of the spontaneously broken $SU(3)_L \times SU(3)_R$ chiral symmetry. To construct the chiral Lagrangian, it is useful to define an H field that transforms under the full $SU(3)_L \times SU(3)_R$ chiral symmetry group in such a way that the transformation reduces to Eq. (5.2) under the unbroken vector subgroup. The transformation of H under $SU(3)_L \times SU(3)_R$ is not uniquely defined, but one can show that all such Lagrangians are related to each other by field redefinitions and so make the same predictions for any physical observable.

For example, one can pick a field $\hat{H}_a$ that transforms as

$$\hat{H}_a \to \hat{H}_b L^\dagger_{ba}, \tag{5.4}$$

under chiral $SU(3)_L \times SU(3)_R$. This transformation property is a little unusual in that it singles out a special role for the $SU(3)_L$ transformations. The parity transform of $\hat{H}$ would then have to transform as in Eq. (5.4) but with L replaced by R. This forces upon us the following choice of parity transformation law:

$$P\hat{H}_a(\mathbf{x}, t)P^{-1} = \gamma^0 \hat{H}_b(-\mathbf{x}, t)\gamma^0 \Sigma_{ba}(-\mathbf{x}, t), \tag{5.5}$$

where Σ is the matrix defined in Eq. (1.99).

Clearly, Eq. (5.4) is not symmetric under $L \leftrightarrow R$, which causes the parity transformation rule to involve the Σ field. It is convenient to have a more symmetrical transformation for H. The key is to introduce a field

$$\xi = \exp(iM/f) = \sqrt{\Sigma}. \tag{5.6}$$

Because of the square root in Eq. (5.6), ξ transforms in a very complicated way under chiral $SU(3)_L \times SU(3)_R$ transformations,

$$\xi \to L\xi U^\dagger = U\xi R^\dagger, \tag{5.7}$$

where U is a function of L, R, and the meson fields $M(x)$. Since it depends on the meson fields, the unitary matrix U is space–time dependent, even though one is making a global chiral transformation with constant values for L and R. Under a $SU(3)_V$ transformation $L = R = V$, ξ has the simple transformation rule

$$\xi \to V\xi V^\dagger, \tag{5.8}$$

and

$$U = V. \tag{5.9}$$

The field

$$H_a = \hat{H}_b \xi_{ba} \tag{5.10}$$

transforms as

$$H_a \to H_b U^\dagger_{ba}, \tag{5.11}$$

under $SU(3)_L \times SU(3)_R$ transformations. Under parity

$$\begin{aligned} PH_aP^{-1} &= \gamma^0 \hat{H}_c \gamma^0 \Sigma_{cb} \xi^\dagger_{ba} \\ &= \gamma^0 H_a \gamma^0, \end{aligned} \tag{5.12}$$

which no longer involves Σ. For a generic matter field, it is convenient to use a field with a transformation law such as Eq. (5.11) that involves U but not L and R, and that reduces to the correct transformation rule under $SU(3)_V$. For example, if X is a matter field that transforms as an adjoint $X \to VXV^\dagger$ under $SU(3)_V$, one would pick the chiral transformation law $X \to UXU^\dagger$.

H and $\hat{H}$ lead to the same predictions for physical observables, since they are related by the field redefinition in Eq. (5.10),

$$H = \hat{H} + \frac{i}{f}\hat{H}M + \cdots, \tag{5.13}$$

which changes off-shell Green's functions but not S-matrix elements. In this chapter we use the H field transforming under $SU(3)_L \times SU(3)_R$ and parity as in Eqs. (5.11) and (5.12). Unless explicitly stated, traces are over the bispinor Lorentz indices and repeated $SU(3)$ indices (denoted by lower-case roman letters) are summed over.

Chiral Lagrangians for matter fields such as H are typically written with ξ rather than Σ for the Goldstone bosons. ξ has a transformation law that involves U, L, and R, whereas the matter field transformation law only involves U. In the construction of invariant Lagrangian terms, it is useful to form combinations of ξ whose transformation laws only involve U. Two such combinations with one derivative are

$$\begin{aligned} \mathbb{V}_\mu &= \frac{i}{2}(\xi^\dagger \partial_\mu \xi + \xi \partial_\mu \xi^\dagger), \\ \mathbb{A}_\mu &= \frac{i}{2}(\xi^\dagger \partial_\mu \xi - \xi \partial_\mu \xi^\dagger), \end{aligned} \tag{5.14}$$

which transform under chiral $SU(3)_L \times SU(3)_R$ transformations as

$$\mathbb{V}_\mu \to U\mathbb{V}_\mu U^\dagger + iU\partial_\mu U^\dagger, \qquad \mathbb{A}_\mu \to U\mathbb{A}_\mu U^\dagger, \tag{5.15}$$

using the transformation rule in Eq. (5.7) for ξ. Thus $\mathbb{A}_\mu$ transforms like the adjoint representation under U and has the quantum numbers of an axial vector field, and $\mathbb{V}_\mu$ transforms like a U-gauge field and has the quantum numbers of a vector field. The $\mathbb{V}_\mu$ field can be used to define a chiral covariant derivative, $D_\mu = \partial_\mu - i\mathbb{V}_\mu$, that can be applied to fields transforming under U. Acting on a field F_a transforming like a **3** the covariant derivative is

$$(DF)_a = \partial F_a - i\mathbb{V}_{ab}F_b, \tag{5.16}$$

and acting on a field G_a transforming like a $\bar{\mathbf{3}}$ the covariant derivative is

$$(DG)_a = \partial G_a + iG_b\mathbb{V}_{ba}. \tag{5.17}$$

The H-field chiral Lagrangian is given by terms that are invariant under chiral $SU(3)_L \times SU(3)_R$ and heavy quark symmetry. The only term with zero derivatives is the H-field mass term $M_H \text{Tr}\, \bar{H}_a H_a$. Scaling the heavy meson fields by $e^{-iM_H v\cdot x}$ removes this mass term. Once this is done, derivatives on the heavy meson fields produce factors of the small residual momentum, and the usual power counting of chiral perturbation theory applies. Scaling the H field to remove the mass term is equivalent to measuring energies in the effective theory relative to the H-field mass M_H, rather than m_Q.

The only allowed terms with one derivative are

$$\mathcal{L} = -i\text{Tr}\, \bar{H}_a v_\mu \left(\partial^\mu \delta_{ab} + i\mathbb{V}^\mu_{ba}\right) H_b + g_\pi \text{Tr}\, \bar{H}_a H_b \gamma_\nu \gamma_5 \mathbb{A}^\nu_{ba}. \tag{5.18}$$

Heavy quark spin symmetry implies that no gamma matrices can occur on the "heavy quark side" of $\bar{H}$ and H in the Lagrangian, i.e., between the two fields in the trace. Any combination of gamma matrices can occur on the light quark side, i.e., to the right of H in the trace. The H fields in Eq. (5.18) are at the same velocity, since low-momentum Goldstone boson exchange does not change the velocity of the heavy quark. It is easy to show that the only gamma matrices that give a nonvanishing contribution to Tr $\bar{H}H\Gamma$ are $\Gamma = 1$ and $\Gamma = \gamma_\mu\gamma_5$. The $\Gamma = 1$ term was the H-field mass term discussed earlier. The $\Gamma = \gamma_\mu\gamma_5$ is the axial coupling to Goldstone bosons. Heavy quark symmetry symmetry implies that at leading order in $1/m_Q$, the coupling constant g_π is independent of the heavy quark mass, i.e., it has the same value for the D and $\bar{B}$ meson systems. The kinetic terms in Lagrange density Eq. (5.18) imply the propagators

$$\frac{i\delta_{ab}}{2(v\cdot k + i\varepsilon)}, \qquad \frac{-i\delta_{ab}(g_{\mu\nu} - v_\mu v_\nu)}{2(v\cdot k + i\varepsilon)}, \tag{5.19}$$

for the P_a and P_a^* mesons, respectively.

The terms in Lagrange density in Eq. (5.18) that arise from $\mathbb{V}_\nu$ contain an even number of pseudo-Goldstone boson fields, whereas the terms that arise from $\mathbb{A}_\nu$, and are proportional to g_π, contain an odd number of pseudo-Goldstone boson fields. Expanding $\mathbb{A}_\nu$, in terms of M, $\mathbb{A}_\nu = -\partial_\nu M/f + \cdots$, gives the P^*PM and P^*P^*M couplings

$$\mathcal{L}_{\text{int}} = \left(\frac{2ig_\pi}{f} P_a^{*\nu\dagger} P_b \partial_\nu M_{ba} + h.c\right) - \frac{2ig_\pi}{f} P_a^{*\alpha\dagger} P_b^{*\beta} \partial^\nu M_{ba} \varepsilon_{\alpha\lambda\beta\nu} v^\lambda. \tag{5.20}$$

The P^*PM and P^*P^*M coupling constants are equal at leading order in $1/m_Q$ as a consequence of heavy quark symmetry; the *PPM* coupling vanishes by parity. The coupling g_π determines the $D^* \to D\pi$ decay width at tree level

$$\Gamma(D^{*+} \to D^0\pi^+) = \frac{g_\pi^2 |\mathbf{p}_\pi|^3}{6\pi f^2}. \tag{5.21}$$

The width for a neutral pion in the final state is one-half of this, by isospin symmetry. The $B^* - B$ mass splitting is less than the pion mass so the analogous $B^* \to B\pi$ decay does not occur.

It is possible to systematically include effects that explicitly break chiral symmetry and heavy quark symmetry as corrections to the chiral Lagrangian. At the order of Λ_{QCD}/m_Q, heavy quark spin symmetry violation occurs only by means of the magnetic moment operator $\bar{Q}_v g \sigma^{\mu\nu} G_{\mu\nu}^A T^A Q_v$, which transforms as a singlet under $SU(3)_L \times SU(3)_R$ chiral symmetry, and as a vector under heavy quark spin symmetry. At leading order in the derivative expansion, its effects are taken into account by adding

$$\delta\mathcal{L}^{(1)} = \frac{\lambda_2}{m_Q} \text{Tr}\, \bar{H}_a \sigma^{\mu\nu} H_a \sigma_{\mu\nu} \tag{5.22}$$

to the Lagrange density in Eq. (5.18). The only effect of $\delta\mathcal{L}^{(1)}$ is to shift the masses of the P and P^* mesons, giving rise to the mass difference

$$\Delta^{(Q)} = m_{P^*} - m_P = -8\frac{\lambda_2}{m_Q}. \tag{5.23}$$

Including this effect, the P and P^* propagators are

$$\frac{i\delta_{ab}}{2\left(v \cdot k + 3\Delta^{(Q)}/4 + i\varepsilon\right)}, \quad \frac{-i\delta_{ab}(g_{\mu\nu} - v_\mu v_\nu)}{2\left(v \cdot k - \Delta^{(Q)}/4 + i\varepsilon\right)}, \tag{5.24}$$

respectively. In the rest frame $v = v_r$, an on-shell P has residual energy $-3\Delta^{(Q)}/4$ and an on-shell P^* has residual energy $\Delta^{(Q)}/4$. It is convenient when dealing with situations in which there is a real P meson and the P^* only appears as a virtual particle to rescale the heavy meson fields by an additional amount, $H \to e^{3i\Delta^{(Q)} v\cdot x/4} H$, so that the P and P^* propagators become

$$\frac{i\delta_{ab}}{2(v \cdot k + i\varepsilon)} \quad \text{and} \quad \frac{-i\delta_{ab}(g_{\mu\nu} - v_\mu v_\nu)}{2\left(v \cdot k - \Delta^{(Q)} + i\varepsilon\right)}, \tag{5.25}$$

respectively. This rescaling is equivalent to measuring energies with respect to the pseudoscalar mass, rather than the average mass of the PP^* multiplet.

Chiral symmetry is explicitly broken by the quark mass matrix m_q, which transforms as $m_q \to L m_q R^\dagger$ under $SU(3)_L \times SU(3)_R$. Chiral symmetry breaking effects at lowest order are given by adding terms linear in m_q to the Lagrange density,

$$\delta\mathcal{L}^{(2)} = \sigma_1 \mathrm{Tr}\, \bar{H}_a H_b (\xi m_q^\dagger \xi + \xi^\dagger m_q \xi^\dagger)_{ab} + \sigma_1' \mathrm{Tr}\, \bar{H}_a H_a (\xi m_q^\dagger \xi + \xi^\dagger m_q \xi^\dagger)_{bb}, \tag{5.26}$$

where m_q is the light quark mass matrix. Expanding ξ in pion fields, $\xi = 1 + \cdots$, it is easy to see that the first term gives rise to mass differences between the heavy mesons due to $SU(3)_V$ breaking. The second term is an overall shift in the meson masses that is due to the light quark masses. It can be distinguished from the chirally symmetric term Tr $\bar{H}H$ because it contains $\pi - H$ interaction terms. The σ_1' term is analogous to the σ term in pion–nucleon scattering. Both terms contain pion interactions of the pseudo-Goldstone bosons with the heavy mesons that do not vanish as the four momenta of the pseudo-Goldstone bosons go to zero, since they contain an explicit factor of chiral symmetry breaking.

The strange quark mass is not as small as the u and d quark masses, and predictions based just on chiral $SU(2)_L \times SU(2)_R$ typically work much better than those that use the full $SU(3)_L \times SU(3)_R$ symmetry group. The results of this section can be used for chiral $SU(2)_L \times SU(2)_R$, by restricting the flavor indices to 1–2, and using the upper 2×2 block of Eq. (1.100) for M, ignoring η. It is important to note that the parameters $g_\pi, \sigma_1, \sigma_1'$, and so on in the $SU(2)_L \times SU(2)_R$ chiral Lagrangian do not have the same values as those in the $SU(3)_L \times SU(3)_R$ chiral Lagrangian. The two-flavor Lagrangian can be obtained from the three-flavor Lagrangian by integrating out the K and η fields.

5.2 g_π in the nonrelativistic constituent quark model

The nonrelativistic constituent quark model is a phenomenological model for QCD in the nonperturbative regime. The quarks in a hadron are treated as nonrelativistic and interact by means of a potential $V(r)$ that is usually fixed to be linear at large distances and Coulombic at short distances. Gluonic degrees of freedom are neglected apart from their implicit role in giving rise to this potential and giving the light quarks their large constituent masses $m_u \simeq m_d \simeq 350\,\mathrm{MeV}$, $m_s \simeq 500\,\mathrm{MeV}$. This simple model predicts many properties of hadrons with surprising accuracy.

We use the quark model to compute the matrix element

$$\langle D^+ | \bar{u}\gamma^3\gamma_5 d | D^{*0} \rangle, \tag{5.27}$$

where the D^{*0} meson has $S_z = 0$ along the spin quantization $\hat{z}$ axis, and the heavy meson states are at rest. To calculate this transition matrix element, we need the operator $\bar{u}\gamma^3\gamma_5 d$ in terms of nonrelativistic constituent quark fields and the D^+ and D^{*0} state vectors. The decomposition of a quark field in terms of nonrelativistic constituent fields is

$$q = \begin{pmatrix} q_{\rm nr}(\uparrow) \\ q_{\rm nr}(\downarrow) \\ -\bar{q}_{\rm nr}(\downarrow) \\ \bar{q}_{\rm nr}(\uparrow) \end{pmatrix} + \cdots, \tag{5.28}$$

where the ellipses denote terms with derivatives. The field $q_{\rm nr}$ destroys a constituent quark and $\bar{q}_{\rm nr}$ creates a constituent antiquark, with spins along the $\hat{z}$ axis as denoted by the arrow. (The lower two elements follow by acting with the charge conjugation operator on the upper two.) Using this decomposition, one finds

$$\begin{aligned} \bar{u}\gamma^3\gamma_5 d = {} & \bar{u}^\dagger_{\rm nr}(\downarrow)\bar{d}_{\rm nr}(\downarrow) - \bar{u}^\dagger_{\rm nr}(\uparrow)\bar{d}_{\rm nr}(\uparrow) \\ & + \text{terms involving quark fields.} \end{aligned} \tag{5.29}$$

In the matrix element Eq. (5.27), the overlap of spatial and color wave functions for the D and D^* states gives unity. The operator only acts nontrivially on the spin-flavor part of the state vector. In our conventions (see Chapter 2), $2i[S^3_Q, D] = -D^{*3}$, and this commutation relation fixes the relative phase of the D and D^* state vectors. Explicitly,

$$\begin{aligned} |D^{*0}\rangle &= |c\uparrow\rangle|\bar{u}\downarrow\rangle + |c\downarrow\rangle|\bar{u}\uparrow\rangle, \\ |D^+\rangle &= i(|c\uparrow\rangle|\bar{d}\downarrow\rangle - |c\downarrow\rangle|\bar{d}\uparrow\rangle). \end{aligned} \tag{5.30}$$

Equations (5.29) and (5.30) yield

$$\langle D^+|\bar{u}\gamma^3\gamma_5 d|D^{*0}\rangle = -2i, \tag{5.31}$$

where the heavy meson states at rest are normalized to two.

The matrix element in Eq. (5.27) can be related to the coupling g_π in the chiral Lagrangian by using the same method that was used in Sec. 1.7 to relate the parameter f to a matrix element of the axial current. Under an infinitesimal axial transformation

$$R = 1 + i\epsilon^B T^B, \qquad L = 1 - i\epsilon^B T^B, \tag{5.32}$$

the QCD Lagrange density changes by

$$\delta\mathcal{L}_{\rm QCD} = -A^B_\mu \partial^\mu \epsilon^B, \tag{5.33}$$

where A_μ^B is the axial current

$$A_\mu^B = \bar{q}\gamma_\mu\gamma_5 T^B q. \tag{5.34}$$

In Eq. (5.34), the $SU(3)$ generator T^B acts on flavor space, and color indices are suppressed. The transformation rule, $\Sigma \to L\Sigma R^\dagger$, implies that under the chiral transformation in Eq. (5.32) the pseudo-Goldstone boson fields transform as

$$\delta M = -f\epsilon^B T^B + \cdots, \tag{5.35}$$

where the ellipses denote terms containing M. Equations (5.4) and (5.10) imply that under an infinitesimal chiral transformation the change in the heavy meson fields vanishes up to terms containing the pseudo-Goldstone boson fields. Consequently the change in the effective chiral Lagrangian Eq. (5.18) under the infinitesimal axial transformation in Eq. (5.32) is

$$\delta\mathcal{L}_{\text{int}} = \left(2g_\pi i P_b^{*\nu} P_a^\dagger T_{ba}^B \partial_\nu \epsilon^B + \text{h.c.}\right) + 2ig_\pi P_a^{*\alpha\dagger} P_b^{*\beta} T_{ba}^B \epsilon_{\alpha\lambda\beta\nu} v^\lambda \partial^\nu \epsilon^B + \cdots. \tag{5.36}$$

Equating $\delta\mathcal{L}_{\text{int}}$ with $\delta\mathcal{L}_{\text{QCD}}$ implies that for matrix elements between heavy meson fields, the axial current can be written as

$$A_\mu^B = \left(-2ig_\pi P_{b\mu}^* P_a^\dagger T_{ba}^B + \text{h.c.}\right) - 2ig_\pi P_a^{*\alpha\dagger} P_b^{*\beta} T_{ba}^B \epsilon_{\alpha\lambda\beta\mu} v^\lambda + \cdots, \tag{5.37}$$

where the ellipses denote terms containing the pseudo-Goldstone boson fields. Using Eq. (5.37) leads to

$$\langle D^+|\bar{u}\gamma^3\gamma_5 d|D^{*0}\rangle = -2ig_\pi, \tag{5.38}$$

and so the nonrelativistic constituent quark model predicts $g_\pi = 1$. Heavy quark flavor symmetry implies that it is the same g_π that determines both the $DD^*\pi$ and $BB^*\pi$ couplings. A similar result for the matrix element of the axial current between nucleons leads to the prediction $g_A = 5/3$ in the nonrelativistic constituent quark model, compared with the experimental value of 1.25. A recent lattice Monte Carlo simulation by the UKQCD Collaboration found $g_\pi = 0.42$ (G.M. de Divitiis et al., hep-lat/9807032).

5.3 $\bar{B} \to \pi e\bar{\nu}_e$ and $D \to \pi\bar{e}\nu_e$ decay

The decay rates for $\bar{B} \to \pi e\bar{\nu}_e$ and $D \to \pi\bar{e}\nu_e$ are determined by the transition matrix elements,

$$\langle\pi(p_\pi)|\bar{q}_a\gamma_\mu(1-\gamma_5)Q\big|P^{(Q)}(p_P)\rangle = f_+^{(Q)}(p_P + p_\pi)_\mu + f_-^{(Q)}(p_P - p_\pi)_\mu. \tag{5.39}$$

Here $f_-^{(Q)}$ can be neglected since its contribution is proportional to the lepton mass in the decay amplitude. The form factors are usually considered to be functions of $q^2 = (p_P - p_\pi)^2$. However, here it is convenient to view the form factors $f_+^{(Q)}$ and $f_-^{(Q)}$ as functions of $v \cdot p_\pi$, where $p_P = m_P v$. The right-hand side of Eq. (5.39) can be rewritten as

$$[f_+^{(Q)} + f_-^{(Q)}] m_P v_\mu + [f_+^{(Q)} - f_-^{(Q)}] p_{\pi\,\mu}. \tag{5.40}$$

In the region of phase space where $v \cdot p_\pi \ll m_Q$, momentum transfers to the light degrees of freedom are small compared with the heavy quark mass, and a transition to HQET is appropriate. Apart from logarithms of m_Q in matching the left-handed current onto the corresponding HQET operator, the left-hand side of Eq. (5.39) depends on m_Q only through the normalization of the $P^{(Q)}$ state, and so it is proportional to $\sqrt{m_Q}$. This gives the following scaling for large m_Q:

$$\begin{aligned} f_+^{(Q)} + f_-^{(Q)} &\sim \mathcal{O}(1/\sqrt{m_Q}), \\ f_+^{(Q)} - f_-^{(Q)} &\sim \mathcal{O}(\sqrt{m_Q}), \end{aligned} \tag{5.41}$$

and in the limit $m_Q \to \infty$, $f_+^{(Q)} = -f_-^{(Q)}$.

Neglecting perturbative corrections, we find the relation between B and D form factors is

$$\begin{aligned} f_+^{(b)} + f_-^{(b)} &= \sqrt{\frac{m_D}{m_B}} [f_+^{(c)} + f_-^{(c)}], \\ f_+^{(b)} - f_-^{(b)} &= \sqrt{\frac{m_B}{m_D}} [f_+^{(c)} - f_-^{(c)}], \end{aligned} \tag{5.42}$$

where in Eqs. (5.42) the form factors for $Q = b$ and $Q = c$ are evaluated at the same value of $v \cdot p_\pi$. Since the decay rate is almost independent of $f_-^{(Q)}$, it is more useful to have a relation just between $f_+^{(b)}$ and $f_+^{(c)}$. Using $f_+^{(Q)} = -f_-^{(Q)}$ in Eq. (5.42) yields such a relation:

$$f_+^{(b)} = \sqrt{\frac{m_B}{m_D}} f_+^{(c)}. \tag{5.43}$$

Equation (5.43) relates the decay rates for $\bar{B} \to \pi e \bar{\nu}_e$ and $D \to \pi \bar{e} \nu_e$ over the part of the phase space where $v \cdot p_\pi \ll m_Q$.

An implicit assumption about the smoothness of the form factors was made in deriving Eq. (5.43). We shall see that this assumption is not valid for very small $v \cdot p_\pi$. In this kinematic region chiral perturbation theory can be used to determine the amplitude.

The operator $\bar{q}_a \gamma^\nu (1 - \gamma_5) Q_v$ transforms as $(\bar{\mathbf{3}}_L, \mathbf{1}_R)$ under $SU(3)_L \times SU(3)_R$ chiral symmetry. This QCD operator is represented in the chiral Lagrangian by an operator constructed out of H and ξ with the same quantum numbers. At

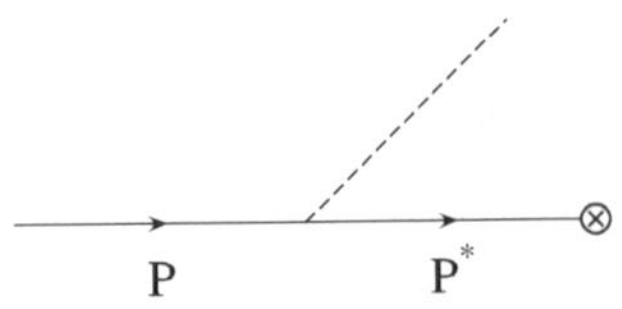

Fig. 5.1. Pole graph contribution to the heavy meson decay form factors. The axial current insertion is denoted by ⊗. The $PP^*\pi$ coupling is from the g_π term in the chiral Lagrangian.

zeroth order in the derivative expansion, it has the form

$$\bar{q}_a\gamma^\nu(1-\gamma_5)Q_v = \frac{a}{2}\text{Tr}\,\gamma^\nu(1-\gamma_5)H_b\,\xi^\dagger_{ba}. \tag{5.44}$$

Heavy quark symmetry has been used to restrict the form of the right-hand side. Operators with derivatives and/or insertions of the light quark mass matrix m_q are higher order in chiral perturbation theory. Recall that $\xi = \exp(iM/f) = 1 + \cdots$, so the part of Eq. (5.44) independent of the pseudo-Goldstone boson fields annihilates P and P^*. This term was already encountered in Sec. 2.8 when we studied the meson decay constants f_D and f_B. At $\mu = m_Q$, $a = \sqrt{m_{P^{(Q)}}}\,f_{P^{(Q)}}$. The part of Eq. (5.44) that is linear in the pseudo-Goldstone boson fields contributes to the $P^{(Q)} \to \pi$ matrix element of Eq. (5.44). There is another contribution from the Feynman diagram in Fig. 5.1 that is also leading order in chiral perturbation theory. Here the $P^{*(Q)}P^{(Q)}\pi$ coupling has one factor of momentum p_π, but that is compensated by the $P^{*(Q)}$ propagator, which is of the order of $1/p_\pi$. The direct and pole contributions together give

$$\begin{aligned} f_+^{(Q)} + f_-^{(Q)} &= \left[\frac{f_{P^{(Q)}}}{f}\right]\left[1 - \frac{g_\pi v\cdot p_\pi}{v\cdot p_\pi + \Delta^{(Q)}}\right], \\ f_+^{(Q)} - f_-^{(Q)} &= \frac{g_\pi f_{P^{(Q)}} m_{P^{(Q)}}}{f\left[v\cdot p_\pi + \Delta^{(Q)}\right]}. \end{aligned} \tag{5.45}$$

Note that $f_+^{(Q)} - f_-^{(Q)}$ is enhanced by $m_{P^{(Q)}}/v\cdot p_\pi$ over $f_+^{(Q)} + f_-^{(Q)}$ and so $f_+^{(Q)} \simeq -f_-^{(Q)}$. Using this relation, we find the prediction of chiral perturbation theory for $f_+^{(Q)}$ becomes

$$f_+^{(Q)} = \frac{g_\pi f_{P^{(Q)}} m_{P^{(Q)}}}{2f\left[v\cdot p_\pi + \Delta^{(Q)}\right]}. \tag{5.46}$$

For $v\cdot p_\pi \gg \Delta^{(b,c)}$ the scaling relation between $f_+^{(b)}$ and $f_+^{(c)}$ in Eqs. (5.43) holds if $1/m_Q$ corrections in the relation between f_B and f_D are small. However, for pions almost at rest, Eq. (5.43) has large corrections because m_π is almost equal to $\Delta^{(c)}$. The derivation of Eq. (5.46) only relies on chiral $SU(2)_L \times SU(2)_R$ symmetry, and it is not necessary to assume that the strange quark mass is also small.

Using chiral $SU(3)_L \times SU(3)_R$, a formula similar to Eq. (5.46) holds for the decay $D \to K\bar{e}\nu_e$. Experimental data on the $D \to K\bar{e}\nu_e$ differential decay rate indicate that $f_+^{(D\to K)}(q^2)$ is consistent with the pole form

$$f_+^{(D\to K)}(q^2) = \frac{f_+^{(D\to K)}(0)}{1 - q^2/M^2}, \tag{5.47}$$

where $M = 2.1$ GeV. With this form for $f_+^{(D\to K)}(q^2)$, the measured decay rate implies that $|V_{cs} f_+^{(D\to K)}(0)| = 0.73 \pm 0.03$. Using $|V_{cs}| = 0.94$, we find this implies that at zero recoil, i.e., $q^2 = q^2_{\rm max} = (m_D - m_K)^2$, the form factor has the value $|f_+^{(D\to K)}(q^2_{\rm max})| = 1.31$. The zero-recoil analog of Eq. (5.46) for this situation is

$$\frac{g_\pi f_{D_s} m_{D_s}}{2f\left(m_K + m_{D_s^*} - m_D\right)} = f_+^{(D\to K)}\left(q^2_{\rm max}\right). \tag{5.48}$$

With the use of the experimental value for $f_+^{(D\to K)}(q^2_{\rm max})$, this implies that $g_\pi f_{D_s} = 129$ MeV. For the lattice value of f_{D_s} in Table 2.3, this gives $g_\pi = 0.6$.

5.4 Radiative D^* decay

The measured branching ratios for D^* decay are presented in Table 5.1. The decay $D^{*0} \to D^+\pi^-$ is forbidden, since $m_{\pi^-} > m_{D^{*0}} - m_{D^+}$. For D^{*0} decay, the electromagnetic and hadronic branching ratios are comparable. Naively, the electromagnetic decay should be suppressed by α compared with the strong one. However, in this case the strong decay is phase space suppressed since $m_{D^*} - m_D$ is very near m_π. For D^{*+} decays, the electromagnetic branching ratio is smaller

Table 5.1. *Measured branching ratios for radiative D^* decay*[a]

Decay mode	Branching ratio(%)
$D^{*0} \to D^0\pi^0$	61.9 ± 2.9
$D^{*0} \to D^0\gamma$	38.1 ± 2.9
$D^{*+} \to D^0\pi^+$	68.3 ± 1.4
$D^{*+} \to D^+\pi^0$	30.6 ± 2.5
$D^{*+} \to D^+\gamma$	1.7 ± 0.5
$D_s^{*+} \to D_s^+\pi^0$	5.8 ± 2.5
$D_s^{*+} \to D_s^+\gamma$	94.2 ± 2.5

[a] The branching ratio for $D^{*+} \to D^+\gamma$ is from a recent CLEO measurement (J. Bartlet et al., Phys. Rev. Lett. 80, 1998, 3919).

than for the D^{*0} case because of a cancellation that we will discuss shortly. The decay $D_s^{*+} \to D_s^+ \pi^0$ is isospin violating and its rate is quite small.

The $D_a^* \to D_a \gamma$ matrix elements have the form [a is an $SU(3)$ index so $D_1 = D^0$, $D_2 = D^+$ and $D_3 = D_s^+$]

$$\mathcal{M}(D_a^* \to D_a \gamma) = e\mu_a \epsilon^{\mu\alpha\beta\lambda} \epsilon_\mu^*(\gamma) v_\alpha k_\beta \epsilon_\lambda(D^*), \tag{5.49}$$

where $\epsilon(\gamma)$ and $\epsilon(D^*)$ are the polarization vectors of the photon and D^*, v is the D^* four velocity (we work in its rest frame where $v = v_r$), and k is the photon's four momentum. The factor $e\mu_a/2$ is a transition magnetic moment. Equation (5.49) yields the decay rate

$$\Gamma(D_a^* \to D_a \gamma) = \frac{\alpha}{3} |\mu_a|^2 |\mathbf{k}|^3. \tag{5.50}$$

The $D_a^* \to D_a \gamma$ matrix elements get contributions from the photon coupling to the light quarks through the light quark part of the electromagnetic current, $\frac{2}{3}\bar{u}\gamma_\mu u - \frac{1}{3}\bar{d}\gamma_\mu d - \frac{1}{3}\bar{s}\gamma_\mu s$, and the photon coupling to the charm quark through its contribution to the electromagnetic current, $\frac{2}{3}\bar{c}\gamma_\mu c$. The part of μ_a that comes from the charm quark part of the electromagnetic current $\mu^{(h)}$ is fixed by heavy quark symmetry. The simplest way to derive it is to examine the $D^* \to D$ matrix element of $\bar{c}\gamma_\mu c$ with the recoil velocity of the D being given approximately by $v' \simeq (1, -\mathbf{k}/m_c)$. Linearizing in $\mathbf{k}$, and using the methods developed in Chapter 2, the heavy quark symmetry prediction for this matrix element is

$$\mu^{(h)} = \frac{2}{3m_c}, \tag{5.51}$$

which is the magnetic moment of a Dirac fermion. Another way to derive Eq. (5.51) is to include electromagnetic interactions in the HQET Lagrangian. Then $\mu^{(h)}$ comes from the order Λ_{QCD}/m_c magnetic moment interaction analgous to the chromagnetic term discussed in Chapter 4. The part of μ_a that comes from the photon coupling to the light quark part of the electromagnetic current is denoted by $\mu_a^{(\ell)}$. It is not fixed by heavy quark symmetry. However, the light quark part of the electromagnetic current transforms as an **8** under the unbroken $SU(3)_V$ group, while the D and D^* states are $\bar{\mathbf{3}}$'s. Since there is only one way to combine a **3** and a $\bar{\mathbf{3}}$ into an **8**, the three transition magnetic moments $\mu_a^{(\ell)}$ are expressible in terms of a single reduced matrix element β,

$$\mu_a^{(\ell)} = Q_a \beta, \tag{5.52}$$

where $Q_1 = 2/3$, $Q_2 = -1/3$ and $Q_3 = -1/3$.

Equation (5.52) is a consequence of $SU(3)_V$ symmetry. Even the relation between $\mu_1^{(\ell)}$ and $\mu_2^{(\ell)}$ depends on $SU(3)_V$ symmetry. The contribution of u and d quarks to the electromagnetic current is a combination of $I = 0$ and $I = 1$ pieces, and so isospin symmetry alone does not imply any relation between $\mu_1^{(\ell)}$ and $\mu_2^{(\ell)}$. We expect that $SU(3)_V$ violations are very important for $\mu_a = \mu^{(h)} + \mu_a^{(\ell)}$. This

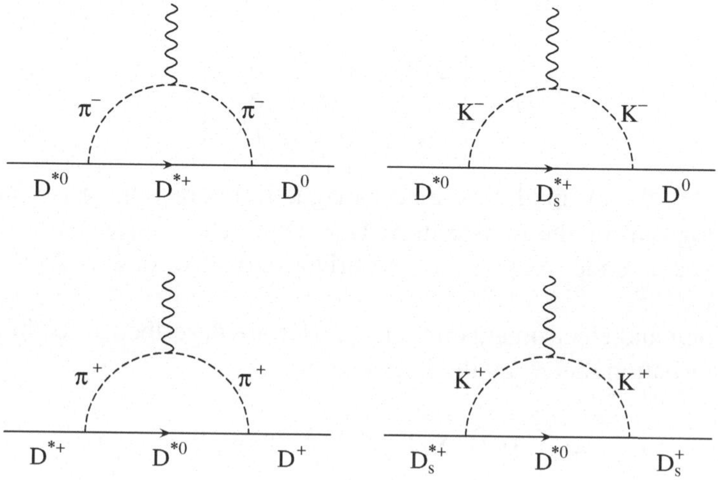

Fig. 5.2. The order $m_q^{1/2}$ corrections to the radiative D^* decay amplitude.

expectation is based on the nonrelativistic constituent quark model. In that model, the $\bar{u}$, $\bar{d}$, and $\bar{s}$ quarks in a D or D^* meson are also treated as heavy, and their contribution to $\mu_a^{(\ell)}$ can be determined in the same way that the charm quark contribution $\mu^{(h)}$ was. This yields

$$\mu_1^{(\ell)} = \frac{2}{3}\frac{1}{m_u}, \quad \mu_2^{(\ell)} = -\frac{1}{3}\frac{1}{m_d}, \quad \mu_3^{(\ell)} = -\frac{1}{3}\frac{1}{m_s}. \tag{5.53}$$

The large $SU(3)_V$ violations occur because for the usual values of the constituent quark masses $m_u \simeq m_d = 350$ MeV, $m_s = 500$ MeV and $m_c = 1.5$ GeV, $\mu_2^{(\ell)}$ and $\mu_3^{(\ell)}$ almost cancel against $\mu^{(h)}$. This cancellation is consistent with the suppression of the $D^{*+} \to D^+\gamma$ rate evident in Table 5.1. With the constituent quark masses given above, the nonrelativistic quark model predictions for the μ_a are $\mu_1 \simeq 2.3$ GeV^{-1}, $\mu_2 = -0.51$ GeV^{-1}, and $\mu_3 = -0.22$ GeV^{-1}.

In chiral perturbation theory the leading $SU(3)_V$ violations are of the order of $m_q^{1/2}$ and come from the Feynman diagrams in Fig. 5.2. The diagrams are calculated with initial and final heavy mesons at the same four velocity v, but the final state D has a residual four momentum $-k$. These diagrams give contributions to μ_a of the order of $m_{(\pi,K)}/f^2$, and their nonanalytic dependence on m_q ensures that higher-order terms in the chiral Lagrangian do not give rise to such terms.

For the Feynman diagrams in Fig. 5.2 to be calculated, the chiral Lagrangian for strong interactions of the pseudo-Goldstone bosons in Eq. (1.102) must be gauged with respect to the electromagnetic subgroup of $SU(3)_V$ transformations. This is done by replacing a derivative of Σ with the covariant derivative

$$\partial_\mu \Sigma \to D_\mu \Sigma = \partial_\mu \Sigma + ie[Q, \Sigma]\mathcal{A}_\mu, \tag{5.54}$$

where

$$Q = \begin{bmatrix} 2/3 & 0 & 0 \\ 0 & -1/3 & 0 \\ 0 & 0 & -1/3 \end{bmatrix}, \tag{5.55}$$

and $\mathcal{A}$ is the photon field. The electromagnetic interactions arise on gauging a $U(1)$ subgroup of the unbroken $SU(3)_V$ symmetry. Since ξ transforms the same way as Σ under $SU(3)_V$, the covariant derivative of ξ is $D_\mu \xi = \partial_\mu \xi + ie[Q, \xi]\mathcal{A}_\mu$.

The strong and electromagnetic interactions are described at leading order in chiral perturbation theory by the Lagrangian

$$\mathcal{L}_{\text{eff}} = \frac{f^2}{8} \text{Tr}\, D^\mu \Sigma (D_\mu \Sigma)^\dagger + v \text{Tr}(m_q \Sigma + m_q \Sigma^\dagger), \tag{5.56}$$

where in this case the trace is over light quark flavor indices. It gives rise to the $MM\gamma$ interaction term

$$\mathcal{L}_{\text{int}} = ie\mathcal{A}_\mu \{[Q, M]_{ab} \partial^\mu M_{ba}\}. \tag{5.57}$$

Using the Feynman rules that follow from Eqs. (5.20) and (5.57), we find the last diagram in Fig. 5.2 gives the following contribution to the $D_s^{*+} \to D_s^+ \gamma$ decay amplitude:

$$\begin{aligned} \delta\mathcal{M} &= i \int \frac{d^n q}{(2\pi)^n} \left(\frac{2}{f} g_\pi \epsilon_{\alpha\lambda\beta\nu} v^\lambda q^\nu \right) \left(\frac{2g_\pi}{f} k^\eta \right) \\ &\quad \times \frac{g^{\alpha\eta}}{2v \cdot q} (e2q_\mu) \left(\frac{1}{q^2 - m_K^2} \right)^2 \epsilon^\beta(D_s^*) \epsilon^\mu(\gamma) \\ &= \frac{4ig^2 e}{f^2} \epsilon_{\alpha\lambda\beta\nu} v^\lambda k^\alpha \epsilon^\beta(D_s^*) \epsilon_\mu^*(\gamma) \int \frac{d^n q}{(2\pi)^n} \frac{q^\nu q^\mu}{\left(q^2 - m_K^2\right)^2 v \cdot q}. \end{aligned} \tag{5.58}$$

In Eq. (5.58) only the linear dependence on k has been kept. The second term in large parentheses is the $D_s^* DK$ coupling. It actually is proportional to $(q - k)^\eta$ but the q^η part does not contribute to $\delta\mathcal{M}$. Similarly, the $KK\gamma$ coupling is proportional to $(2q - k)_\mu$, but the k_μ part is omitted in Eq. (5.58), since it does not contribute to $\delta\mathcal{M}$. Finally, the part proportional to $v^\alpha v^\eta$ in the D_s^* propagator also does not contribute to $\delta\mathcal{M}$ and is not displayed in Eq. (5.58).

Combining denominators using Eq. (3.6) gives

$$\begin{aligned} \delta\mathcal{M} &= \frac{16ig_\pi^2 e}{f^2} \epsilon_{\alpha\lambda\beta\nu} v^\lambda k^\alpha \epsilon^\beta(D_s^*) \epsilon_\mu^*(\gamma) \\ &\quad \times \int_0^\infty d\lambda \int \frac{d^n q}{(2\pi)^n} \frac{q^\nu q^\mu}{\left(q^2 + 2\lambda v \cdot q - m_K^2\right)^3}. \end{aligned} \tag{5.59}$$

Shifting the integration variable q by λv, we find this becomes

$$\delta\mathcal{M} = \frac{16ig_\pi^2 e}{nf^2}\epsilon_{\alpha\lambda\beta\mu}v^\lambda k^\alpha \epsilon^\beta(D_s^*)\epsilon^{*\mu}(\gamma)\int_0^\infty d\lambda \int \frac{d^n q}{(2\pi)^n}\frac{q^2}{\left(q^2 - m_K^2 - \lambda^2\right)^3}. \tag{5.60}$$

Consequently, the contribution of this Feynman diagram to the transition magnetic moment is

$$\delta\mu_3^{(\ell)} = \frac{16ig_\pi^2}{nf^2}\int_0^\infty d\lambda \int \frac{d^n q}{(2\pi)^n}\frac{q^2}{\left(q^2 - m_K^2 - \lambda^2\right)^3}. \tag{5.61}$$

Performing the q integration using Eq. (1.44) yields

$$\delta\mu_3^{(\ell)} = -\frac{4g_\pi^2\Gamma(2-n/2)}{f^2 2^n \pi^{n/2}}\int_0^\infty d\lambda\left(\lambda^2 + m_K^2\right)^{-2+n/2}. \tag{5.62}$$

Using Eq. (3.11), we find it easy to see that the integral over λ is proportional to $\epsilon = 4 - n$ and so the expression for $\delta\mu_3^{(\ell)}$ is finite as $\epsilon \to 0$. Taking this limit, we find

$$\delta\mu_3^{(\ell)} = \frac{g_\pi^2 m_K^2}{2\pi^2 f^2}\int_0^\infty \frac{d\lambda}{\left(\lambda^2 + m_K^2\right)} = \frac{g_\pi^2 m_K}{4\pi f^2}. \tag{5.63}$$

A similar calculation can be done for the other diagrams. Identifying f with f_K for the kaon loops, and with f_π for the pion loops, we have

$$\begin{aligned}
\mu_1^{(\ell)} &= \frac{2}{3}\beta - \frac{g_\pi^2 m_K}{4\pi f_K^2} - \frac{g_\pi^2 m_\pi}{4\pi f_\pi^2},\\
\mu_2^{(\ell)} &= -\frac{1}{3}\beta + \frac{g_\pi^2 m_\pi}{4\pi f_\pi^2},\\
\mu_3^{(\ell)} &= -\frac{1}{3}\beta + \frac{g_\pi^2 m_K}{4\pi f_K^2}.
\end{aligned} \tag{5.64}$$

Using f_K for kaon loops and f_π for pion loops reduces somewhat the magnitude of the kaon loops compared with the pion loops. Experience with kaon loops in chiral perturbation theory for interactions of the pseudo-Goldstone bosons with nucleons suggests that such a suppression is present.

It remains to consider the isospin violating decay $D_s^{*+} \to D_s^+\pi^0$. The two sources of isospin violation are electromagnetic interactions and the difference between the d and u quark masses, $m_d - m_u$. In chiral perturbation theory the pole type diagram in Fig. 5.3 dominates the part of the amplitude coming from the quark mass difference. The $\eta - \pi^0$ mixing is given in Eq. (1.104). Using this

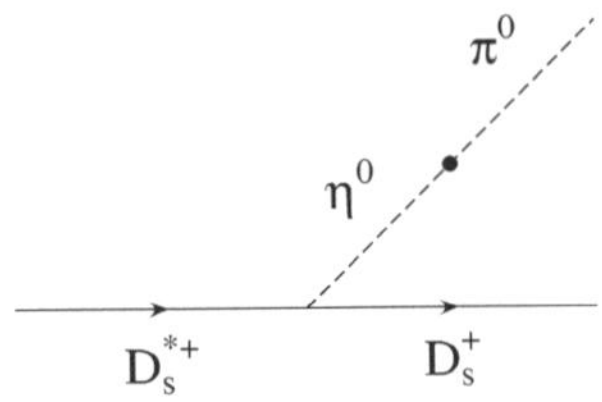

Fig. 5.3. Leading contribution to the isospin violating decay $D_s^* \to D_s \pi^0$.

and Eq. (5.20), we find the decay rate is

$$\Gamma(D_s^{*+} \to D_s^+ \pi^0) = \frac{g_\pi^2}{48\pi f^2} \left[\frac{m_d - m_u}{m_s - (m_u + m_d)/2}\right]^2 |\mathbf{p}_\pi|^3. \qquad (5.65)$$

The measured mass difference $m_{D_s^*} - m_{D_s} = 144.22 \pm 0.60$ MeV implies that $|\mathbf{p}_\pi| \simeq 49.0$ MeV. In chiral perturbation theory, this is the dominant contribution coming from the quark mass difference because it is suppressed by only $(m_d - m_u)/m_s \simeq 1/43.7$, as opposed to $(m_d - m_u)/4\pi f$. The isospin violating electromagnetic contribution is expected to be less important since α/π is smaller than $(m_d - m_u)/m_s$.

The measured branching ratios in Table 5.1 determine the values of g_π and β. There are two solutions, since one has to solve a quadratic equation. Using the above results gives either ($g_\pi = 0.56$, $\beta = 3.5$ GeV^{-1}) or ($g_\pi = 0.24$, $\beta = 0.85$ GeV^{-1}). In evaluating these parameters, we have set $f = f_\pi$ for the hadronic modes. The values obtained for g_π are smaller than the quark model prediction discussed in Sec. 5.2. Of course there is a large uncertainty in this determination of g_π, since the experimental errors on branching ratios for the isospin violating decay $D_s^{*+} \to D_s^+ \pi^0$ and the radiative decay $D^{*+} \to D^+ \gamma$ are large, and because higher-order terms in chiral perturbation theory that have been neglected may be important.

5.5 Chiral corrections to $\bar{B} \to D^{(*)} e \bar{\nu}_e$ form factors

In Chapter 4, the nonperturbative order $\Lambda_{\rm QCD}/m_Q$, corrections to B decay form factors, such as the semileptonic $\bar{B} \to D^{(*)} e \bar{\nu}_e$ form factors, $h_\pm(w)$, $h_V(w)$ and $h_{A_j}(w)$, were discussed. It seems reasonable that nonperturbative corrections to the form factors should be expandable in powers of $(\Lambda_{\rm QCD}/m_Q)$, since the Lagrangian has an expansion in inverse powers of the heavy quark mass m_Q. However, because of the small values for the u and d quark masses, this ends up not being the case because of pole and loop diagrams involving pions. This point is illustrated below with the help of two examples: $\bar{B} \to D^* \pi e \bar{\nu}_e$, which has pole terms, and $\bar{B} \to D e \bar{\nu}_e$, which has pion loop terms.

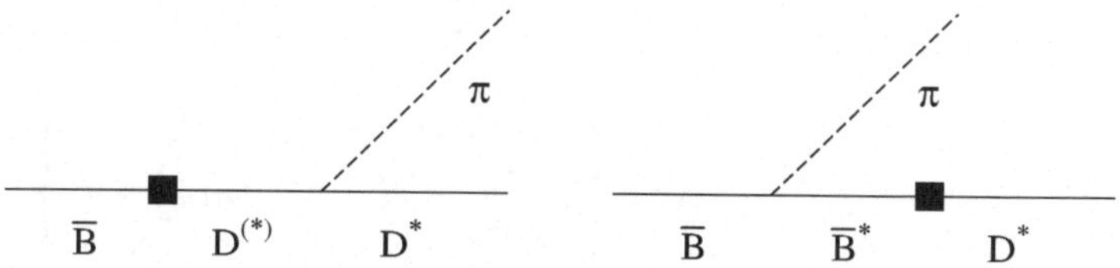

Fig. 5.4. Pole diagram contribution to the $\bar{B} \to D^* \pi e \bar{\nu}_e$ form factors. The solid box is an insertion of the axial current, Eq. (5.66).

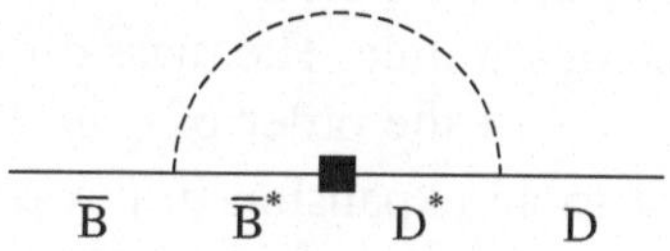

Fig. 5.5. One-loop correction to the $\bar{B} \to D e \bar{\nu}_e$ form factors.

The weak current $\bar{c}\gamma_\mu(1-\gamma_5)b$ is a singlet under chiral $SU(3)_L \times SU(3)_R$ transformations. At leading order in chiral perturbation theory, this operator is represented in the chiral Lagrangian by

$$\bar{c}\gamma_\mu(1-\gamma_5)b = -\xi(w)\mathrm{Tr}\,\bar{H}^{(c)}_{av'}\gamma_\mu(1-\gamma_5)H^{(b)}_{av}, \tag{5.66}$$

where we have now put back the heavy quark and velocity labels. $\xi(w)$ is the Isgur-Wise function.

Equation (5.66) contains no powers of the pion fields. This implies that at leading order in chiral perturbation theory the $\bar{B} \to D^* \pi e \bar{\nu}_e$ amplitudes come from pole diagrams in Fig. 5.4. The propagator for the intermediate D meson is

$$\frac{i}{p_\pi \cdot v + \Delta^{(c)}} \tag{5.67}$$

where p_π is the pion momentum and $\Delta^{(c)}$ is the D^*–D mass difference, which is of the order of $\Lambda^2_{\mathrm{QCD}}/m_c$. Clearly, the form factors for this decay depend on $\Delta^{(c)}/v \cdot p_\pi$, and so do not simply have an expansion in $\Lambda_{\mathrm{QCD}}/m_Q$. A similar conclusion holds for the $\bar{B} \to \pi e \bar{\nu}_e$ form factors discussed in Sec. 5.3.

To compute $\bar{B} \to D e \bar{\nu}_e$ form factors, one needs the $\bar{B} \to D$ matrix element of Eq. (5.66). The leading order of chiral perturbation theory is the tree-level matrix element of this operator. At higher order in chiral perturbation theory one needs loop diagrams, as well as additional terms in Eq. (5.66) involving derivatives and insertions of the light quark mass matrix. At one loop the diagram in Fig. 5.5 contributes to the form factors for $\bar{B} \to D e \bar{\nu}_e$ decay. This contribution is proportional to $g_\pi^2/(4\pi f)^2$ and depends also on the pion mass, m_π, and the $D^* - D$ mass difference, $\Delta^{(c)}$ (here, for simplicity, we neglect the $B^* - B$ mass difference). At zero recoil, Fig. 5.5, the wave-function renormalization diagrams and a tree-level contribution from an order $1/m_c^2$ operator give the contribution

$$\delta h_+(1) = -\frac{3g_\pi^2}{32\pi^2 f^2}\Delta^{(c)2}\left\{\ln\frac{\mu^2}{m_\pi^2} + F\left[\Delta^{(c)}/m_\pi\right] + C\right\}, \tag{5.68}$$

where μ is the scale parameter of dimensional regularization, and F is a dimensionless function that can be computed by explicitly evaluating the diagrams. Here C is the contribution of the local order $1/m_c^2$ operator. Any dependence on μ in a Feynman diagram is logarithmic. The mass difference $\Delta^{(c)}$ is of the order of $1/m_c$, and the pion mass is of the order of $\sqrt{m_q}$. Expanding F in a power series in $\Delta^{(c)}$ is equivalent to an expansion in powers of $1/m_c$. Expanding in powers of $\Delta^{(c)}$ gives

$$F = -\frac{3\pi}{4}\frac{\Delta^{(c)}}{m_\pi} + \frac{6}{5}\frac{\Delta^{(c)2}}{m_\pi^2} + \cdots. \tag{5.69}$$

Dimensional analysis dictates that for terms in $\delta h_+(1)$ of the order of $[\Delta^{(c)}]^n \sim (1/m_c)^n$, $n=3,4\ldots$, the coefficients have the form $1/m_\pi^{n-2}$ and diverge as $m_\pi \to 0$. Nonperturbative corrections to the form factor $h_+(1)$ are not suppressed by powers of $(\Lambda_{\rm QCD}/m_c)$ but are much larger, of the order of $\Lambda_{\rm QCD}^{3n/2+2}/m_c^{n+2}\, {m_q}^{n/2}$ for $n \geq 0$. Note that in accordance with Luke's theorem there is no order $1/m_c$ term in $\delta h_+(1)$.

The heavy quark limit is m_c large, and the chiral limit is m_q small. Expanding F in powers of $\Delta^{(c)}$ is equivalent to taking the heavy quark limit where m_c is large while keeping m_q fixed. If one first takes the chiral limit where m_q is small while keeping m_c fixed, one should instead expand in powers of m_π. This expansion has the form

$$F = \left[\frac{2}{3} - \ln\frac{4\Delta^{(c)2}}{m_\pi^2}\right] + \frac{m_\pi^2}{\Delta^{(c)2}}\left[\frac{9}{2} - \frac{3}{2}\ln\frac{4\Delta^{(c)2}}{m_\pi^2}\right] - 2\pi\frac{m_\pi^3}{\Delta^{(c)3}} + \cdots, \tag{5.70}$$

with coefficients with positive powers of m_c in the higher order terms, which diverge as $m_c \to \infty$.

While the expansions in Eqs. (5.69) and (5.70) have divergent coefficients in the $m_\pi \to 0$ and $m_c \to \infty$ limits, respectively, the contribution of Fig. 5.5 and the wave-function renormalization diagrams to $h_+(1)$ is perfectly well defined. The chiral Lagrangian for heavy mesons can be used as long as $\Delta^{(c)}$ and m_π are both much smaller than $\Lambda_{\rm CSB}$, and the correction Eq. (5.68) is smaller than unity, irrespective of the value for the ratio $\Delta^{(c)}/m_\pi$. The interesting terms which cause divergent coefficients arise because of the ratio of two small scales, m_π and $\Delta^{(c)}$, and all such effects are computable in chiral perturbation theory.

5.6 Problems

1. Compute the magnetic moments of the baryon octet in the nonrelativistic constituent quark model, and compare the results with the experimental data.

2. Neglecting the u and d quark masses, show that in chiral perturbation theory

$$\frac{f_{B_s}}{f_B} = 1 - \frac{5}{6}\left(1 + 3g_\pi^2\right)\frac{m_K^2}{16\pi^2 f^2}\left(\ln\frac{m_K^2}{\mu^2} + C\right) + \cdots .$$

C is a constant and the ellipsis denotes terms of higher order in chiral perturbation theory. The $\ln(m_K^2/\mu^2)$ term is referred to as a "chiral logarithm." The μ dependence of this term is canceled by a corresponding μ dependence in the coefficient C. If m_K were extremely small, the logarithm would dominate over the constant C.

3. The form factors for $D \to K\pi\bar{e}\nu_e$ are defined by

$$\begin{aligned}\langle\pi(p_\pi)K(p_K)|\bar{s}\gamma_\mu P_L c|D(p_D)\rangle &= i\omega_+ P_\mu + i\omega_- Q_\mu \\ &\quad + ir(p_D - P)_\mu + h\epsilon_{\mu\alpha\beta\gamma}p_D^\alpha P^\beta Q^\gamma,\end{aligned}$$

where

$$P = p_K + p_\pi, \qquad Q = p_K - p_\pi.$$

Use chiral perturbation theory to express the form factors $\omega_\pm$, r and h for $D^+ \to K^-\pi^+\bar{e}\nu_e$ in terms of f_D, f, g_π, $\Delta^{(c)} = m_{D^*} - m_D$ and $\mu_s = m_{D_s} - m_D$.

4. Verify Eqs. (5.64) and (5.65).

5. Evaluate $F(\Delta/m_\pi)$ in Eq. (5.68). Expand in $1/m_c$ and m_π and verify Eqs. (5.69) and (5.70).

6. The low-lying baryons containing a heavy quark Q transforms as a **6** and $\bar{\mathbf{3}}$ under $SU(3)_V$. Under the full chiral $SU(3)_L \times SU(3)_R$ the fields that destroy these baryons transform as

$$S_{ab}^\mu \to U_{ac}U_{bd}S_{cd}^\mu, \qquad T_a \to T_a U_{ab}^\dagger,$$

where (see Problem 10 in Chapter 2)

$$S_{ab}^\mu = \frac{1}{\sqrt{3}}(\gamma_\mu + v_\mu)\gamma_5 B_{ab} + B_{ab}^{*\mu}.$$

Velocity and heavy quark labels are suppressed here.

(a) In the case $Q = c$ identify the various components of the fields T_a, B_{ab}, and $B_{ab}^{*\mu}$ with baryon states in Table 2.1.

(b) Argue that at leading order in $1/m_Q$, m_q, and derivatives, the chiral Lagrangian for heavy baryon pseudo-Goldstone boson interactions is

$$\begin{aligned}\mathcal{L} = &-i\bar{S}_{ab}^\mu(v\cdot D)S_{\mu ab} + \Delta M\bar{S}_{ab}^\mu S_{\mu ab} + i\bar{T}_a(v\cdot D)T_a + ig_2\epsilon_{\mu\nu\sigma\lambda}\bar{S}_{ab}^\mu v^\nu S_{cb}^\lambda \mathbb{A}_{ac}^\sigma \\ &+ g_3\left(\epsilon_{abc}\bar{T}_a S_{\mu cd}\mathbb{A}_{bd}^\mu + \text{h.c.}\right).\end{aligned}$$

Define how the covariant derivative D acts on S_{ab}^μ and T_a.

5.7 References

The implications of combining heavy quark and chiral symmetries were first studied in:

G. Burdman and J.F. Donoghue, Phys. Lett. B280 (1992) 287
M.B. Wise, Phys. Rev. D45 (1992) 2188
T.-M. Yan, H.-Y. Cheng, C.-Y. Cheung, G.-L. Lin, Y.C. Lin, and H.-L. Yu, Phys. Rev. D46 (1992) 1148 [erratum: ibid D55 (1997) 5851]

Some applications were studied in:

B. Grinstein, E. Jenkins, A.V. Manohar, M.J. Savage, and M.B. Wise, Nucl. Phys. B380 (1992) 369
J.L. Goity, Phys. Rev. D46 (1992) 3929
H.-Y. Cheng, C.-Y. Cheung, G.-L. Lin, Y.C. Lin, T.-M. Yan, and H.-L. Yu, Phys. Rev. D47 (1993) 1030, D49 (1994) 2490, D49 (1994) 5857 [erratum: ibid D55 (1997) 5851]
J.F. Amundson, C.G. Boyd, E. Jenkins, M.E. Luke, A.V. Manohar, J.L. Rosner, M.J. Savage, and M.B. Wise, Phys. Lett. B296 (1992) 415
P. Cho and H. Georgi, Phys. Lett. B296 (1992) 408 [erratum: ibid B300 (1993) 410]
C.G. Boyd and B. Grinstein, Nucl. Phys. B442 (1995) 205
H.-Y. Cheng, C.-Y. Cheung, W. Dimm, G.-L. Lin, Y.C. Lin, T.-M. Yan, and H.-L. Yu, Phys. Rev. D48 (1993) 3204
E. Jenkins, Nucl. Phys. B412 (1994) 181
I. Stewart, Nucl. Phys. B529 (1998) 62
G. Burdman, Z. Ligeti, M. Neubert, and Y. Nir, Phys. Rev. D49 (1994) 2331

For discussions of excited charm meson decays using chiral perturbation theory see:

U. Kilian, J. Korner, and D. Pirjol, Phys. Lett. B288 (1992) 360
A.F. Falk and M.E. Luke, Phys. Lett. B292 (1992) 119
J. Korner, D. Pirjol, and K. Schilcher, Phys. Rev. D47 (1993) 3955
P. Cho and S. Trivedi, Phys. Rev. D50 (1994) 381
A.F. Falk and T. Mehen, Phys. Rev. D53 (1996) 231

The isospin violating decay $D_s^* \to D_s \pi$ was discussed in:

P. Cho and M.B. Wise, Phys. Rev. D49 (1994) 6228

Chiral corrections for $B \to D^{(*)}$ were discussed in:

E. Jenkins and M.J. Savage, Phys. Lett. B281 (1992) 331
L. Randall and M.B. Wise, Phys. Lett. B303 (1993) 135

For a review, see:

R. Casalbuoni, A. Deandrea, N. Di Bartolomeo, R. Gatto, F. Feruglio, and G. Nardulli, Phys. Rep. 281 (1997) 145

Chiral perturbation theory for baryons containing a heavy quark was introduced in:

P. Cho, Phys. Lett. B285 (1992) 145

and some applications appear in:

M.J. Savage, Phys. Lett. B359 (1995) 189
M. Lu, M.J. Savage, and J. Walden, Phys. Lett. B365 (1996) 244

6
Inclusive weak decay

In this chapter, we will study inclusive weak decays of hadrons containing a b quark. The lowest mass meson or baryon containing a b quark decays weakly, since the strong and electromagnetic interactions preserve quark flavor. One of the main results of this chapter is the demonstration that the parton model picture that inclusive heavy hadron decay is the same as free heavy quark decay is exact in the $m_b \to \infty$ limit. In addition, we will show how to include radiative and nonperturbative corrections to the leading-order formula in a systematic way. The analysis closely parallels that of deep inelastic scattering in Sec. 1.8.

6.1 Inclusive semileptonic decay kinematics

Semileptonic $\bar{B}$-meson decays to final states containing a charm quark arise from matrix elements of the weak Hamiltonian density

$$H_W = \frac{4G_F}{\sqrt{2}} V_{cb}\, \bar{c}\gamma^\mu P_L b\, \bar{e}\gamma_\mu P_L \nu_e. \tag{6.1}$$

In exclusive three-body decays such as $\bar{B} \to De\bar{\nu}_e$, one looks at the decay into a definite final state, such as $De\bar{\nu}_e$. The differential decay distribution has two independent kinematic variables, which can be chosen to be E_e and E_{ν_e}, the energy of the electron and antineutrino. The decay distribution depends implicitly on the masses of the initial and final particles, which are constants. In inclusive decays, one ignores all details about the final hadronic state X_c and sums over all final states containing a c quark. Here X_c can be a single-particle state, such as a D meson, or a multiparticle state, such as $D\pi$. In addition to the usual two kinematic variables E_e and E_{ν_e} for exclusive semileptonic decays, there is an additional kinematic variable in $\bar{B} \to X_c e\bar{\nu}_e$ decay since the invariant mass of the final hadronic system can vary. The third variable will be chosen to be q^2, the invariant mass of the virtual W boson. The diagrams for semileptonic b-quark and $\bar{B}$-meson decays are shown in Fig. 6.1.

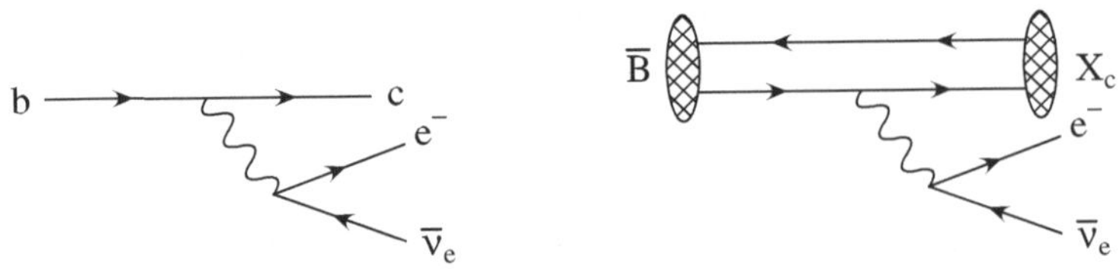

Fig. 6.1. Weak decay diagrams for semileptonic quark and hadron decay.

In the $\bar{B}$ rest frame, the differential decay distribution for inclusive semileptonic decay is

$$\begin{aligned}\frac{d\Gamma}{dq^2\, dE_e\, dE_{\nu_e}}\\ &= \int \frac{d^4 p_e}{(2\pi)^4} \int \frac{d^4 p_{\nu_e}}{(2\pi)^4}\, 2\pi\delta(p_e^2) 2\pi\delta(p_{\nu_e}^2)\theta(p_e^0)\theta(p_{\nu_e}^0)\\ &\quad\times \delta(E_e - p_e^0)\delta(E_{\nu_e} - p_{\nu_e}^0)\delta\left[q^2 - (p_e + p_{\nu_e})^2\right]\\ &\quad\times \sum_{X_c} \sum_{\substack{\text{lepton}\\ \text{spins}}} \frac{|\langle X_c e \bar{\nu}_e | H_W | \bar{B}\rangle|^2}{2m_B} (2\pi)^4 \delta^4\left[p_B - (p_e + p_{\nu_e}) - p_{X_c}\right], \end{aligned} \tag{6.2}$$

where we have used the familiar formula $d^3p/(2E) = d^4 p\delta(p^2 - m^2)\theta(p^0)$ and neglected the electron mass. The phase space integrations can be performed in the rest frame of the $\bar{B}$ meson. After summation over final hadronic states X_c, the only relevant angle is that between the electron and the neutrino three momenta. Nothing depends on the direction of the neutrino momentum, and integrating over it gives a factor of 4π. One can then choose the z axis for the electron momentum to be aligned along the neutrino direction. Integrating over the electron azimuthal angle gives a factor of 2π. Consequently, the lepton phase space is

$$d^3 p_e d^3 p_{\nu_e} = 8\pi^2 |\mathbf{p}_e|^2 d|\mathbf{p}_e| |\mathbf{p}_{\nu_e}|^2 d|\mathbf{p}_{\nu_e}| d\cos\theta, \tag{6.3}$$

where θ is the angle between the electron and neutrino directions. The three remaining integrations are fixed by the three delta functions. Using $\delta(p_e^2) = \delta(E_e^2 - |\mathbf{p}_e|^2)$ to perform the integration over $|\mathbf{p}_e|$, $\delta(p_{\nu_e}^2) = \delta(E_{\nu_e}^2 - |\mathbf{p}_{\nu_e}|^2)$ to perform the integration over $|\mathbf{p}_{\nu_e}|$, and $\delta[q^2 - (p_e + p_{\nu_e})^2] = \delta[q^2 - 2E_e E_{\nu_e}(1 - \cos\theta)]$ to perform the integration over $\cos\theta$ gives

$$\frac{d\Gamma}{dq^2\, dE_e\, dE_{\nu_e}} = \frac{1}{4} \sum_{X_c} \sum_{\substack{\text{lepton}\\ \text{spins}}} \frac{|\langle X e \bar{\nu}_e | H_W | \bar{B}\rangle|^2}{2m_B} \delta^4\left[p_B - (p_e + p_{\nu_e}) - p_{X_c}\right]. \tag{6.4}$$

The weak matrix element in Eq. (6.4) can be factored into a leptonic matrix element and a hadronic matrix element, since leptons do not have any strong interactions. Corrections to this result are suppressed by powers of G_F or α, and they arise from radiative corrections due to additional electroweak gauge bosons propagating between the quark and lepton lines. The matrix element average is conventionally written as the product of hadronic and leptonic tensors,

$$\frac{1}{4}\sum_{X_c}\sum_{\substack{\text{lepton}\\ \text{spins}}} \frac{|\langle X_c e\bar{\nu}_e|H_W|\bar{B}\rangle|^2}{2m_B}(2\pi)^3\delta^4\big[p_B - (p_e + p_{\nu_e}) - p_{X_c}\big]$$
$$= 2G_F^2|V_{cb}|^2 W_{\alpha\beta}L^{\alpha\beta}, \tag{6.5}$$

where the leptonic tensor is

$$L^{\alpha\beta} = 2\left(p_e^\alpha p_{\nu_e}^\beta + p_e^\beta p_{\nu_e}^\alpha - g^{\alpha\beta}p_e\cdot p_{\nu_e} - i\epsilon^{\eta\beta\lambda\alpha}p_{e\eta}p_{\nu_e\lambda}\right) \tag{6.6}$$

and the hadronic tensor is defined by

$$W^{\alpha\beta} = \sum_{X_c}(2\pi)^3\delta^4(p_B - q - p_{X_c})\frac{1}{2m_B}$$
$$\times\langle\bar{B}(p_B)|J_L^{\dagger\alpha}|X_c(p_{X_c})\rangle\langle X_c(p_{X_c})|J_L^\beta|\bar{B}(p_B)\rangle, \tag{6.7}$$

with $J_L^\alpha = \bar{c}\gamma^\alpha P_L b$, the left-handed current. In Eq. (6.7), $q = p_e + p_{\nu_e}$ is the sum of electron and antineutrino four momenta. Here $W_{\alpha\beta}$ is a second-rank tensor that depends on $p_B = m_B v$ and q, the momentum transfer to the hadronic system. The relation $p_B = m_B v$ defines v as the four velocity of the $\bar{B}$ meson. The b quark can have a small three velocity of the order of $1/m_b$ in the $\bar{B}$-meson rest frame, and this effect is included in the $1/m_b$ corrections computed later in this chapter.

The most general tensor $W_{\alpha\beta}$ is

$$W_{\alpha\beta} = -g_{\alpha\beta}W_1 + v_\alpha v_\beta W_2 - i\epsilon_{\alpha\beta\mu\nu}v^\mu q^\nu W_3 + q_\alpha q_\beta W_4 + (v_\alpha q_\beta + v_\beta q_\alpha)W_5. \tag{6.8}$$

The scalar structure functions W_j are functions of the Lorentz invariant quantities q^2 and $q\cdot v$. Using Eqs. (6.8), (6.6), and (6.5), we find the differential cross section in Eq. (6.4) becomes

$$\frac{d\Gamma}{dq^2\,dE_e\,dE_{\nu_e}} = \frac{G_F^2|V_{cb}|^2}{2\pi^3}\big[W_1q^2 + W_2(2E_eE_{\nu_e} - q^2/2)$$
$$+ W_3q^2(E_e - E_{\nu_e})\big]\theta(4E_eE_{\nu_e} - q^2), \tag{6.9}$$

where we have explicitly included the θ function that sets the lower limit for the E_{ν_e} integration because it will play an important role later in this chapter. The functions W_4 and W_5 do not contribute to the decay rate, since $q_\alpha L^{\alpha\beta} = q_\beta L^{\alpha\beta} = 0$ in the limit that the electron mass is neglected. These terms have to be included in decays to the τ.

The neutrino is not observed, and so one integrates the above expression over E_{ν_e} to get the differential spectrum $d\Gamma/dq^2\,dE_e$. For a fixed electron energy the minimum value of q^2 occurs when the electron and neutrino are parallel (i.e.,

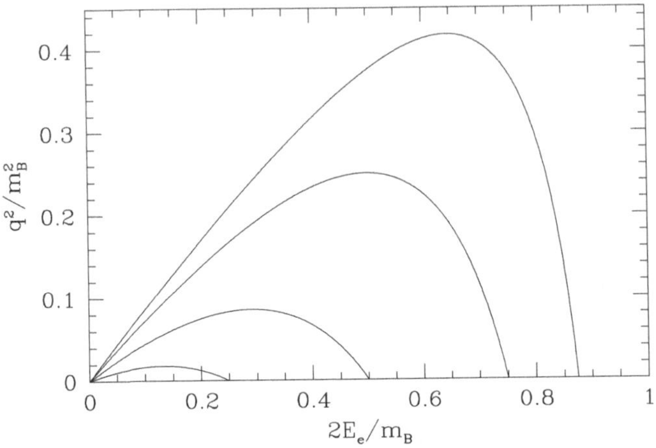

Fig. 6.2. The allowed q^2 values as a function of the electron energy E_e, for different values of the final state hadronic mass m_{X_c}. The entire region inside the curve is allowed. The curves are (from the outermost curve in) for $m_{X_c} = m_D$, $(m_{X_c}/m_B)^2 = 0.25, 0.5,$ and 0.75, respectively.

$\cos\theta = 1$), and the maximum value occurs when the electron and neutrino are antiparallel (i.e., $\cos\theta = -1$). Hence

$$0 < q^2 < \frac{2E_e}{(m_B - 2E_e)}\left(m_B^2 - 2E_e m_B - m_{X_c^{\min}}^2\right), \tag{6.10}$$

where $X_c^{\min}$ is the lowest mass state containing a charm quark, i.e., the D meson. The maximum electron energy is

$$E_e^{\max} = \frac{m_B^2 - m_{X_c^{\min}}^2}{2m_B}, \tag{6.11}$$

which occurs at $q^2 = 0$. The allowed q^2 values as a function of E_e are plotted in Fig. 6.2. For a given value of the final hadronic system mass m_{X_c}, electron energy E_e, and q^2, the neutrino energy E_{ν_e} is

$$E_{\nu_e} = \left(\frac{m_B^2 - m_{X_c}^2 + q^2}{2m_B}\right) - E_e. \tag{6.12}$$

Consequently, integrating $\mathrm{d}\Gamma/\mathrm{d}q^2\,\mathrm{d}E_e\,\mathrm{d}E_{\nu_e}$ over E_{ν_e} (at fixed q^2 and E_e) to get $\mathrm{d}\Gamma/\mathrm{d}q^2\,\mathrm{d}E_e$ is equivalent to averaging over a range of final-state hadronic masses. We will see later in this chapter that in some regions of phase space, the validity of the operator product expansion for inclusive decays depends on hadronic mass averaging. For values of q^2 and E_e near the boundary of the allowed kinematic region, $q^2(m_B - 2E_e) - 2E_e(m_B^2 - 2E_e m_B - m_{X_c^{\min}}^2) = 0$, only final hadronic states with masses near $m_{X_c^{\min}}$ get averaged over in the integration over E_{ν_e}.

The hadronic tensor $W_{\alpha\beta}$ parameterizes all strong interaction physics relevant for inclusive semileptonic $\bar{B}$ decay. It can be related to the discontinuity of a time-ordered product of currents across a cut. Consider the time-ordered product

$$T_{\alpha\beta} = -i \int d^4x e^{-iq\cdot x} \frac{\langle \bar{B}|T\left[J^\dagger_{L\alpha}(x)\, J_{L\beta}(0)\right]|\bar{B}\rangle}{2m_B}. \tag{6.13}$$

Inserting a complete set of states between the currents in each time ordering, using the analogs of Eqs. (1.159), applying the identity

$$\theta(x^0) = -\frac{1}{2\pi i}\int_{-\infty}^{\infty} d\omega\, \frac{e^{-i\omega x^0}}{\omega + i\varepsilon}, \tag{6.14}$$

and performing the integration over d^4x gives, in the $\bar{B}$ rest frame,

$$\begin{aligned} T_{\alpha\beta} &= \sum_{X_c} \frac{\langle \bar{B}|J^\dagger_{L\alpha}|X_c\rangle\langle X_c|J_{L\beta}|\bar{B}\rangle}{2m_B(m_B - E_X - q^0 + i\varepsilon)}(2\pi)^3\delta^3(\mathbf{q}+\mathbf{p}_X) \\ &\quad - \sum_{X_{\bar{c}bb}} \frac{\langle \bar{B}|J_{L\beta}|X_{\bar{c}bb}\rangle\langle X_{\bar{c}bb}|J^\dagger_{L\alpha}|\bar{B}\rangle}{2m_B(E_X - m_B - q^0 - i\varepsilon)}(2\pi)^3\delta^3(\mathbf{q}-\mathbf{p}_X). \end{aligned} \tag{6.15}$$

Here X_c is a complete set of hadronic states containing a c quark, and $X_{\bar{c}bb}$ is a complete set of hadronic states containing two b quarks and a $\bar{c}$ quark. At fixed $\mathbf{q}$ the time-ordered product of currents $T_{\alpha\beta}$ has cuts in the complex q^0 plane along the real axis. One cut is in the region $-\infty < q^0 < m_B - \sqrt{m^2_{X_c^{\min}} + |\mathbf{q}|^2}$, and the other cut is in the region $\infty > q^0 > \sqrt{m^2_{X_{\bar{c}bb}^{\min}} + |\mathbf{q}|^2} - m_B$. The imaginary part of T (i.e., the discontinuity across the cut) can be evaluated using

$$\frac{1}{\omega + i\varepsilon} = P\frac{1}{\omega} - i\pi\delta(\omega), \tag{6.16}$$

where P denotes the principal value. This gives

$$\begin{aligned} \frac{1}{\pi}\,\text{Im}\, T_{\alpha\beta} &= -\sum_{X_c} \frac{\langle \bar{B}|J^\dagger_{L\alpha}|X_c\rangle\langle X_c|J_{L\beta}|\bar{B}\rangle}{2m_B}(2\pi)^3\delta^4(p_B - q - p_X) \\ &\quad - \sum_{X_{\bar{c}bb}} \frac{\langle \bar{B}|J_{L\beta}|X_{\bar{c}bb}\rangle\langle X_{\bar{c}bb}|J^\dagger_{L\alpha}|\bar{B}\rangle}{2m_B}(2\pi)^3\delta^4(p_B + q - p_X). \end{aligned} \tag{6.17}$$

The first of these two terms is just $-W_{\alpha\beta}$. For values of q and p_B in semileptonic $\bar{B}$ decay, the argument of the δ function in the second term of Eq. (6.17) is never zero, and it does not contribute to the imaginary part of T. It is convenient to express $T_{\alpha\beta}$ in terms of Lorentz scalar structure functions just as we did for $W_{\alpha\beta}$:

$$T_{\alpha\beta} = -g_{\alpha\beta}T_1 + v_\alpha v_\beta T_2 - i\epsilon_{\alpha\beta\mu\nu}v^\mu q^\nu T_3 + q_\alpha q_\beta T_4 + (v_\alpha q_\beta + v_\beta q_\alpha)T_5. \tag{6.18}$$

The T_j's are functions of q^2 and $q \cdot v$. One can study T_j in the complex $q \cdot v$ plane for fixed q^2. This is a Lorentz invariant way of studying the analytic structure discussed above. For the cut associated with physical hadronic states containing a c quark, $(p_B - q) - p_X = 0$, which implies that $v \cdot q = (m_B^2 + q^2 - m_{X_c}^2)/2m_B$. This cut is in the region $-\infty < v \cdot q < (m_B^2 + q^2 - m_{X_c^{\min}}^2)/2m_B$ (see Fig. 6.3). In contrast, the cut corresponding to physical hadronic states with a $\bar{c}$ quark and two b quarks has $(p_B + q) - p_X = 0$, which implies that $v \cdot q = (m_{X_{\bar{c}bb}}^2 - m_B^2 - q^2)/2m_B$. This cut occurs in the region $(m_{X_{\bar{c}bb}^{\min}}^2 - m_B^2 - q^2)/2m_B < v \cdot q < \infty$. These cuts are widely separated for all values of q^2 allowed in $\bar{B} \to X_c e \bar{\nu}_e$ semileptonic decay, $0 < q^2 < (m_B - m_{X_c^{\min}})^2$. The minimum separation between the cuts occurs for the maximal value of q^2. Approximating hadron masses by that of the heavy quark they contain (e.g., $m_{X_c^{\min}} = m_c$, $m_{X_{\bar{c}bb}^{\min}} = m_c + 2m_b$, etc.), we find the minimum separation between the two cuts is, $4m_c$, which is much greater than the scale $\Lambda_{\rm QCD}$ of nonperturbative strong interactions. The discontinuity across the left-hand cut gives the structure functions for inclusive semileptonic decay:

$$-\frac{1}{\pi}\,\mathrm{Im}\,T_j = W_j \quad \text{(left-hand cut only)}. \tag{6.19}$$

The double differential decay rate $\mathrm{d}\Gamma/\mathrm{d}q^2\,\mathrm{d}E_e$ can be obtained from the triple differential rate $\mathrm{d}\Gamma/\mathrm{d}q^2\,\mathrm{d}E_e\,\mathrm{d}E_{\nu_e}$, or equivalently, $\mathrm{d}\Gamma/\mathrm{d}q^2\,\mathrm{d}E_e\,\mathrm{d}v \cdot q$, by integrating over $q \cdot v = E_e + E_{\nu_e}$. Integrals of the structure functions $W_j(q^2, v \cdot q)$ over $v \cdot q$ are then related to integrals of T_j over the contour C shown in Fig. 6.3.

The situation is similar for $b \to u$ decays. The results for this case can be obtained from our previous discussion just by changing the subscript c to u. However, since the u quark mass is negligible, the separation between the two cuts is not large compared with the scale of the strong interactions, $\Lambda_{\rm QCD}$, when q^2 is near its maximal value for $b \to u$ decays. The significance of this will be commented on later in this chapter.

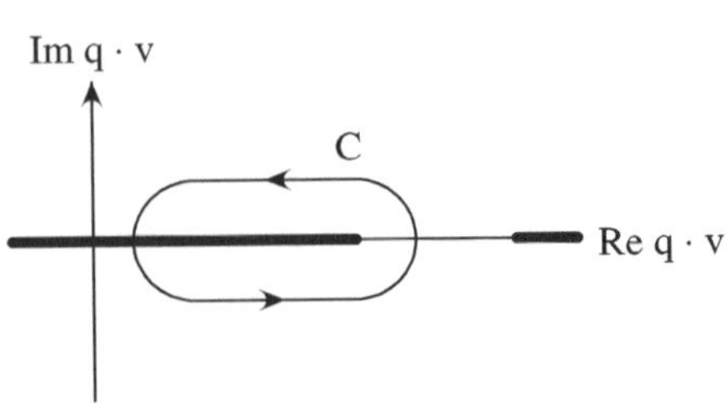

Fig. 6.3. Contour for the T_j integral.

6.2 The operator product expansion

The structure functions T_j can be expressed in terms of matrix elements of local operators using the operator product expansion to simplify the time-ordered product of currents,

$$-i \int d^4x e^{-iq\cdot x} T\left[J^\dagger_{L\alpha}(x)\, J_{L\beta}(0)\right], \tag{6.20}$$

whose $\bar{B}$-meson matrix element is $T_{\alpha\beta}$. The coefficients of the operators that occur in this expansion can be reliably computed by using QCD perturbation theory, in any region of $v \cdot q$ that is far away (compared with $\Lambda_{\rm QCD}$) from the cuts. We compute the coefficients of the operators that occur in the operator product expansion by using quark and gluon matrix elements of Eq. (6.20). These operators will involve the b-quark field, covariant derivatives D, and the gluon field strength $G^A_{\mu\nu}$. At dimension six and above, the light quark fields also occur.

At lowest order in perturbation theory the matrix element of Eq. (6.20) between b-quark states with momentum $m_b v + k$ is (see Fig. 6.4)

$$\frac{1}{(m_b v - q + k)^2 - m_c^2 + i\varepsilon} \bar{u}\gamma_\alpha P_L (m_b \not{v} - \not{q} + \not{k})\gamma_\beta P_L u. \tag{6.21}$$

In the matrix elements of interest, q is usually of the order of m_b, but k is of the order of $\Lambda_{\rm QCD}$. Expanding in powers of k gives an expansion in powers of $\Lambda_{\rm QCD}/m_b$, and thus an expansion in $1/m_b$ of the form factors T_j.

6.2.1 Lowest order

The order k^0 terms in the expansion of Eq. (6.21) are

$$\begin{aligned} &\frac{1}{\Delta_0}\bar{u}[(m_b v - q)_\alpha \gamma_\beta + (m_b v - q)_\beta \gamma_\alpha - (m_b \not{v} - \not{q}) g_{\alpha\beta} \\ &\quad - i\epsilon_{\alpha\beta\lambda\eta}(m_b v - q)^\lambda \gamma^\eta] P_L u, \end{aligned} \tag{6.22}$$

where

$$\Delta_0 = (m_b v - q)^2 - m_c^2 + i\varepsilon, \tag{6.23}$$

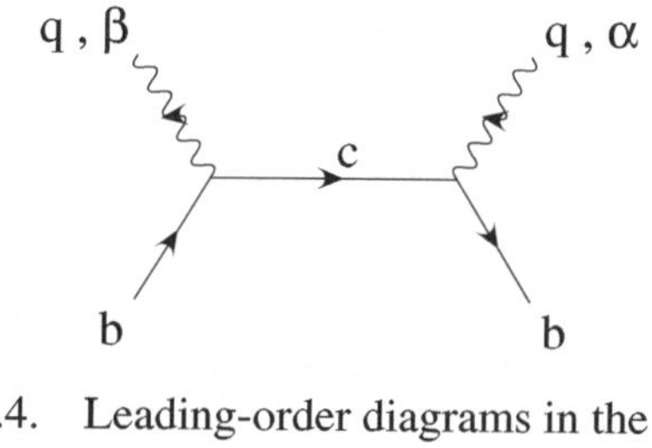

Fig. 6.4. Leading-order diagrams in the OPE.

and we have used the identity in Eq. (1.119). The matrix elements of the dimension-three operators $\bar{b}\gamma^\lambda b$ and $\bar{b}\gamma^\lambda\gamma_5 b$ between b-quark states are $\bar{u}\gamma^\lambda u$ and $\bar{u}\gamma^\lambda\gamma_5 u$, respectively, so the operator product expansion is obtained by replacing u and $\bar{u}$ in Eq. (6.22) by the fields b and $\bar{b}$, respectively. Finally, to get the T_j we take the hadronic matrix elements of the operators,

$$\langle \bar{B}(p_B)|\bar{b}\gamma_\lambda b|\bar{B}(p_B)\rangle = 2p_{B\lambda} = 2m_B v_\lambda \tag{6.24}$$

and

$$\langle \bar{B}(p_B)|\bar{b}\gamma_\lambda\gamma_5 b|\bar{B}(p_B)\rangle = 0. \tag{6.25}$$

The latter matrix element vanishes because of the parity invariance of the strong interactions. Equation (6.24) follows because $\bar{b}\gamma_\lambda b$ is the conserved b-quark number current. The b-quark number charge $Q_b = \int d^3x\bar{b}\gamma_0 b$ acts on $\bar{B}$-meson states as $Q_b|\bar{B}\rangle = |\bar{B}\rangle$, since they have unit b-quark number. Note that Eqs. (6.24) and (6.25) are exact. There are no corrections of order $\Lambda_{\rm QCD}/m_b$ to these relations and hence at this level in the OPE there is no need to make a transition to the heavy quark effective theory.

The T_j's that follow from Eqs. (6.24), (6.25), and (6.22) are

$$\begin{aligned} T_1^{(0)} &= \frac{1}{2\Delta_0}(m_b - q\cdot v), \\ T_2^{(0)} &= \frac{1}{\Delta_0}m_b, \\ T_3^{(0)} &= \frac{1}{2\Delta_0}. \end{aligned} \tag{6.26}$$

At this level in the operator product expansion, the entire cut reduces to a simple pole. The W_j's that follow from Eq. (6.26) are

$$\begin{aligned} W_1^{(0)} &= \frac{1}{4}\left(1 - \frac{q\cdot v}{m_b}\right)\delta\left[v\cdot q - \left(\frac{q^2 + m_b^2 - m_c^2}{2m_b}\right)\right], \\ W_2^{(0)} &= \frac{1}{2}\delta\left[v\cdot q - \left(\frac{q^2 + m_b^2 - m_c^2}{2m_b}\right)\right], \\ W_3^{0} &= \frac{1}{4m_b}\delta\left[v\cdot q - \left(\frac{q^2 + m_b^2 - m_c^2}{2m_b}\right)\right]. \end{aligned} \tag{6.27}$$

Putting these expressions into Eq. (6.9) and performing the integration over neutrino energies using the δ function in Eq. (6.27) gives

$$\frac{d\Gamma}{d\hat{q}^2\,dy} = \frac{G_F^2|V_{cb}|^2 m_b^5}{192\pi^3}12(y - \hat{q}^2)(1 + \hat{q}^2 - \rho - y)\theta(z), \tag{6.28}$$

where

$$y = 2E_e/m_b, \qquad \hat{q}^2 = q^2/m_b^2, \qquad \rho = m_c^2/m_b^2, \tag{6.29}$$

and

$$z = 1 + \hat{q}^2 - \rho - \hat{q}^2/y - y \tag{6.30}$$

are convenient dimensionless variables. This is the same result one obtains from calculating the decay of a free b quark. Integrating over $\hat{q}^2$ gives the lepton energy spectrum

$$\frac{\mathrm{d}\Gamma}{\mathrm{d}y} = \frac{G_F^2 |V_{cb}|^2 m_b^5}{192\pi^3}\left[2(3-2y)y^2 - 6y^2\rho - \frac{6y^2\rho^2}{(1-y)^2} + \frac{2(3-y)y^2\rho^3}{(1-y)^3}\right], \tag{6.31}$$

which also is the same as obtained from free quark decay. Including perturbative QCD corrections to the coefficient of the operator $\bar{b}\gamma_\lambda b$ in the operator product expansion would reproduce the perturbative QCD corrections to the b-quark decay rate.

At linear order in k, Eq. (6.21) contains the terms

$$\begin{aligned}
&\frac{1}{\Delta_0}\bar{u}(k_\alpha\gamma_\beta + k_\beta\gamma_\alpha - g_{\alpha\beta}\not{k} - i\epsilon_{\alpha\beta\lambda\eta}k^\lambda\gamma^\eta)P_L u \\
&\quad - \frac{2k\cdot(m_b v - q)}{\Delta_0^2}\bar{u}[(m_b v - q)_\alpha\gamma_\beta + (m_b v - q)_\beta\gamma_\alpha \\
&\quad - (m_b\not{v} - \not{q})g_{\alpha\beta} - i\epsilon_{\alpha\beta\lambda\eta}(m_b v - q)^\lambda\gamma^\eta]P_L u.
\end{aligned} \tag{6.32}$$

These produce terms in the operator product expansion of the form $\bar{b}\gamma_\lambda(iD_\tau - m_b v_\tau)b$ and $\bar{b}\gamma_\lambda\gamma_5(iD_\tau - m_b v_\tau)b$. Converting the b-quark fields in QCD to those in the heavy quark effective theory gives, at leading order in $1/m_b$, the operators $\bar{b}_v\gamma_\lambda iD_\tau b_v = v_\lambda\bar{b}_v iD_\tau b_v$ and $\bar{b}_v\gamma_\lambda\gamma_5 iD_\tau b_v$. The second of these has a vanishing $\bar{B}$-meson matrix element by parity invariance of the strong interactions. The first has a matrix element that can be written in the form

$$\langle\bar{B}(v)|\bar{b}_v iD_\tau b_v|\bar{B}(v)\rangle = X v_\tau. \tag{6.33}$$

Contracting both sides with v^τ, we find that the equation of motion in HQET, $(iv\cdot D)b_v = 0$, implies that $X = 0$. There are no matrix elements of dimension-four operators that occur in the OPE for the T_j's. This means that, when the differential semileptonic $\bar{B}$-meson decay rate is expressed in terms of the bottom and charm quark masses, there are no corrections suppressed by a single power of $\Lambda_{\mathrm{QCD}}/m_b$.

6.2.2 Dimension-five operators

There are several sources of contributions from dimension-five operators to the operator product expansion. At order k^1 we found in the previous subsection that the operators $\bar{b}\gamma_\lambda(iD_\tau - m_b v_\tau)b$ and $\bar{b}\gamma_\lambda\gamma_5(iD_\tau - m_b v_\tau)b$ occur. Including

$1/m_b$ corrections to the relationship between QCD and HQET operators gives rise to dimension-five operators in HQET. Recall from Chapter 4 that at order $1/m_b$, the relationship between the b-quark field in QCD and in HQET is (to zeroth order in α_s)

$$b(x) = e^{-im_b v\cdot x}\left(1 + \frac{i\not{D}}{2m_b}\right) b_v(x), \tag{6.34}$$

and the order $1/m_b$ HQET Lagrange density is

$$\mathcal{L}_1 = -\bar{b}_v \frac{D^2}{2m_b} b_v - \bar{b}_v g \frac{G_{\alpha\beta}\sigma^{\alpha\beta}}{4m_b} b_v. \tag{6.35}$$

As was noted in Chapter 4, one can drop the $\perp$ subscript on D at this order. Equations (6.34) and (6.35) imply that at order $1/m_b$ (and zeroth order in α_s),

$$\begin{aligned}\bar{b}\gamma_\lambda(iD_\tau - m_b v_\tau)b &= \bar{b}_v\gamma_\lambda iD_\tau b_v + i\int d^4x\, T[\bar{b}_v\gamma_\lambda iD_\tau b_v(0)\,\mathcal{L}_1(x)]\\ &+ \bar{b}_v\left(\frac{-i\overleftarrow{\not{D}}}{2m_b}\right)\gamma_\lambda iD_\tau b_v + \bar{b}_v\gamma_\lambda iD_\tau \frac{i\not{D}}{2m_b} b_v.\end{aligned} \tag{6.36}$$

Equation (6.36) is an operator matching condition. The matrix element of the left-hand side is to be taken in QCD, and of the right-hand side in HQET between hadrons states constructed using the lowest order Lagrangian. The effects of the $1/m_b$ corrections to the Lagrangian have been explicitly included as a time-ordered product term in the operator. Equation (6.36) is valid at a subtraction point $\mu = m_b$, with corrections of order $\alpha_s(m_b)$.

Let us consider the $\bar{B}$-matrix element of the various terms that occur on the right-hand side of Eq. (6.36). We have already shown that the equations of motion of HQET imply that $\bar{b}_v\gamma_\lambda iD_\tau b_v$ has zero $\bar{B}$-meson matrix elements. For the time-ordered product, we note that γ_λ can be replaced by v_λ and write

$$\langle\bar{B}(v)|i\int d^4x\, T[\bar{b}_v iD_\tau b_v(0)\,\mathcal{L}_1(x)]|\bar{B}(v)\rangle = A v_\tau. \tag{6.37}$$

Contracting with v_τ yields

$$\langle\bar{B}(v)|i\int d^4x T[\bar{b}_v(iv\cdot D)b_v(0)\,\mathcal{L}_1(x)]|\bar{B}(v)\rangle = A. \tag{6.38}$$

At tree level, the time-ordered product is evaluated by using $(v\cdot D)S_h(x-y) = \delta^4(x-y)$, where S_h is the HQET propagator. Consequently,

$$A = -\langle\bar{B}(v)|\mathcal{L}_1(0)|\bar{B}(v)\rangle = -\frac{\lambda_1}{m_b} - \frac{3\lambda_2}{m_b}, \tag{6.39}$$

where λ_1 and λ_2 were defined in Eqs. (4.23). There is another way to evaluate the $\bar{B}$-matrix element of the first two terms on the right-hand side of Eq. (6.36). Instead of including the time-ordered product, one evaluates the matrix element of the first term by using the equations of motion that include $\mathcal{O}(1/m_Q)$ terms in the Lagrangian, i.e., $\bar{b}_v\,(iv\cdot D)\,b_v = -\mathcal{L}_1$.

Using the operator identity $[D_\alpha, D_\beta] = igG_{\alpha\beta}$, we find the last two terms on the right side of Eq. (6.36) become

$$\bar{b}_v \frac{i\not{D}}{2m_b}\gamma_\lambda iD_\tau b_v + \bar{b}_v\gamma_\lambda iD_\tau \frac{i\not{D}}{2m_b} b_v = \bar{b}_v \frac{iD_{(\lambda}iD_{\tau)}}{m_b} b_v - \bar{b}_v g \frac{G_{\alpha\tau}\sigma^\alpha{}_\lambda}{2m_b} b_v, \tag{6.40}$$

where parentheses around indices denote that they are symmetrized, i.e.,

$$a^{(\alpha}b^{\beta)} = \frac{1}{2}(a^\alpha b^\beta + a^\beta b^\alpha).$$

For the operator with symmetrized covariant derivatives we write

$$\langle \bar{B}(v)|\bar{b}_v iD_{(\lambda}iD_{\tau)}b_v|\bar{B}(v)\rangle = Y(g_{\lambda\tau} - v_\lambda v_\tau). \tag{6.41}$$

The tensor structure on the right-hand side of this equation follows from the HQET equation of motion $(iv\cdot D)b_v = 0$, which implies that it must vanish when either index is contracted with the b quark's four velocity. To fix Y we contract both sides with $g^{\lambda\tau}$, giving

$$Y = \frac{1}{3}\langle \bar{B}(v)|\bar{b}_v(iD)^2 b_v|\bar{B}(v)\rangle = \frac{2}{3}\lambda_1. \tag{6.42}$$

Finally we need

$$\langle \bar{B}(v)|\bar{b}_v gG_{\alpha\tau}\sigma^\alpha{}_\lambda b_v|\bar{B}(v)\rangle = Z(g_{\lambda\tau} - v_\lambda v_\tau), \tag{6.43}$$

where again the tensor structure on the right-hand side follows from the fact that contracting v^λ into it must vanish, since $\bar{b}_v\sigma^\alpha{}_\lambda v^\lambda b_v = 0$. Contracting both sides of Eq. (6.43) with the metric tensor yields

$$Z = \frac{1}{3}\langle \bar{B}(v)|\bar{b}_v gG_{\alpha\beta}\sigma^{\alpha\beta}b_v|\bar{B}(v)\rangle = -4\lambda_2. \tag{6.44}$$

Combining these results we have that the order k^1 terms in Eq. (6.32) give the following contribution to the T_j's:

$$\begin{aligned}
T_1^{(1)} &= -\frac{1}{2m_b}(\lambda_1 + 3\lambda_2)\left\{\frac{1}{6\Delta_0} - \frac{(m_b - q\cdot v)^2}{\Delta_0^2} + \frac{2}{3}\frac{[q^2 - (q\cdot v)^2]}{\Delta_0^2}\right\},\\
T_2^{(1)} &= -\frac{1}{2m_b}(\lambda_1 + 3\lambda_2)\left[\frac{5}{3\Delta_0} - \frac{2m_b(m_b - v\cdot q)}{\Delta_0^2} + \frac{4}{3}\frac{m_b v\cdot q}{\Delta_0^2}\right],\\
T_3^{(1)} &= \frac{1}{2m_b}(\lambda_1 + 3\lambda_2)\frac{5}{3}\left(\frac{m_b - v\cdot q}{\Delta_0^2}\right).
\end{aligned} \tag{6.45}$$

6.2.3 Second order

The order k^2 terms in Eq. (6.21) are

$$-2\frac{k\cdot(m_bv-q)}{\Delta_0^2}\bar{u}(k_\alpha\gamma_\beta+k_\beta\gamma_\alpha-g_{\alpha\beta}\not{k}-i\epsilon_{\alpha\beta\lambda\eta}k^\lambda\gamma^\eta)P_Lu$$
$$+\left\{\frac{4[k\cdot(m_bv-q)]^2}{\Delta_0^3}-\frac{k^2}{\Delta_0^2}\right\}\bar{u}[(m_bv-q)_\alpha\gamma_\beta+(m_bv-q)_\beta\gamma_\alpha$$
$$-(m_b\not{v}-\not{q})g_{\alpha\beta}-i\epsilon_{\alpha\beta\lambda\eta}(m_bv-q)^\lambda\gamma^\eta]P_Lu. \tag{6.46}$$

These can be expressed in terms of matrix elements of the operators $\bar{b}\gamma^\lambda(iD-m_bv)^{(\alpha}(iD-m_bv)^{\beta)}b$ and $\bar{b}\gamma^\lambda\gamma_5(iD-m_bv)^{(\alpha}(iD-m_bv)^{\beta)}b$. The operator involving γ_5 will not contribute to $\bar{B}$-meson matrix elements by parity. Rewriting the result using HQET operators, we find the only operator that occurs is $v^\lambda\bar{b}_v iD^{(\alpha}iD^{\beta)}b_v$. Its matrix element is given by Eqs. (6.41) and (6.42). So we find that the terms with two k's give the following contribution to the structure functions:

$$T_1^{(2)}=\frac{1}{6}\lambda_1(m_b-v\cdot q)\left\{\frac{4}{\Delta_0^3}[q^2-(v\cdot q)^2]-\frac{3}{\Delta_0^2}\right\},$$
$$T_2^{(2)}=\frac{1}{3}\lambda_1 m_b\left\{\frac{4}{\Delta_0^3}[q^2-(v\cdot q)^2]-\frac{3}{\Delta_0^2}-\frac{2v\cdot q}{m_b\Delta_0^2}\right\}, \tag{6.47}$$
$$T_3^{(2)}=\frac{1}{6}\lambda_1\left\{\frac{4}{\Delta_0^3}[q^2-(v\cdot q)^2]-\frac{5}{\Delta_0^2}\right\}.$$

At zeroth order in α_s, the b-quark matrix element of the operator $\bar{b}\sigma_{\alpha\beta}G^{\alpha\beta}b$ vanishes. To find the part of the operator product expansion proportional to this operator, we need to consider the $b\to b+$ gluon matrix element of the time-ordered product. At tree level it is given by the Feynman diagram in Fig. 6.5. The matrix element has the initial b quark with residual momentum $p/2$, a final b quark with residual momentum $-p/2$, and the gluon with outgoing momentum p. This choice is convenient since the denominators of the c-quark propagators do not contribute to the p dependence at linear order in p. The part of this Feynman diagram with no factors of the gluon four momentum, p, is from the $b\to b+$gluon matrix element of operators we have already found, with the gluon

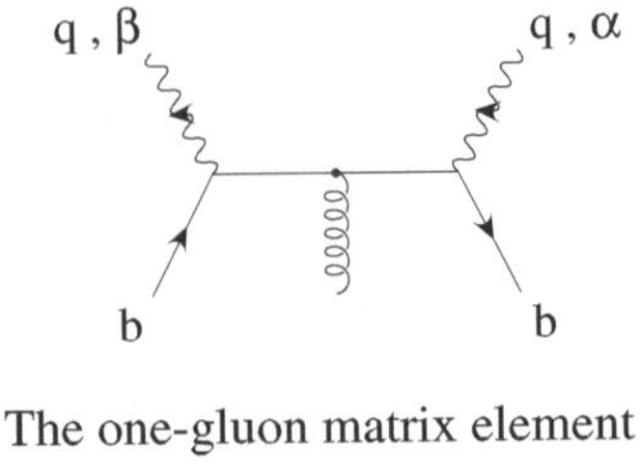

Fig. 6.5. The one-gluon matrix element in the OPE.

field coming from the covariant derivative $D = \partial + igA$. The part linear in p is

$$gT^A \varepsilon^{A\lambda *}(p) \frac{1}{2\Delta_0^2} \bar{u} \gamma_\alpha [-\not{p} \gamma_\lambda (m_b \not{v} - \not{q}) + (m_b \not{v} - \not{q}) \gamma_\lambda \not{p}] \gamma_\beta P_L u, \qquad (6.48)$$

where $\varepsilon^{A\lambda}$ is the gluon polarization vector. Only the part of this antisymmetric under interchange $p \leftrightarrow \varepsilon^*$ contributes to the operator we are considering. Equation (1.119) is used to reexpress the product of three-gamma matrices in the square brackets of Eq. (6.48) in terms of a single-gamma matrix. Only the part proportional to the Levi-Civita tensor survives. Applying the identity of Eq. (1.119) one more time shows that the term linear in p is reproduced by the matrix element of the operator

$$\frac{g}{2\Delta_0^2} \bar{b} G_{\mu\nu} \epsilon^{\mu\nu\lambda\sigma} (m_b v - q)_\lambda (g_{\alpha\sigma} \gamma_\beta + g_{\beta\sigma} \gamma_\alpha - g_{\alpha\beta} \gamma_\sigma + i \epsilon_{\alpha\sigma\beta\tau} \gamma^\tau \gamma_5) P_L b. \qquad (6.49)$$

Here we have used the replacement

$$p^\beta T^A \varepsilon^{A\lambda *} \to -\frac{i}{2} G^{\beta\lambda}$$

for the part antisymmetric in β and λ.

The transition to HQET is made by replacing b-quark fields in the above by b_v. The operators that occur are $\bar{b}_v G^{\mu\nu} \gamma^\lambda \gamma_5 b_v$ and $\bar{b}_v G^{\mu\nu} \gamma^\lambda b_v$. Because of the antisymmetry on the indices μ and ν, parity invariance of the strong interaction forces the latter operator to have a zero matrix element between $\bar{B}$-meson states. The matrix element of the other operator can be written as

$$\langle \bar{B}(v) | \bar{b}_v g G^{\mu\nu} \gamma^\lambda \gamma_5 b_v | \bar{B}(v) \rangle = N \epsilon^{\mu\nu\lambda\tau} v_\tau. \qquad (6.50)$$

Contracting both sides of this equation with $\epsilon_{\mu\nu\lambda\rho} v^\rho$ and using the identity

$$\epsilon_{\mu\nu\lambda\rho} v^\rho \bar{b}_v \gamma^\lambda \gamma_5 b_v = -\bar{b}_v \sigma_{\mu\nu} b_v \qquad (6.51)$$

yields

$$N = -2\lambda_2. \qquad (6.52)$$

Consequently, the $b \to b +$ gluon matrix element gives these additional contributions to the structure functions:

$$\begin{aligned} T_1^{(g)} &= \lambda_2 \frac{(m_b - v \cdot q)}{2\Delta_0^2}, \\ T_2^{(g)} &= -\lambda_2 \frac{m_b}{\Delta_0^2}, \\ T_3^{(g)} &= \lambda_2 \frac{1}{2\Delta_0^2}. \end{aligned} \qquad (6.53)$$

Summing the three contributions we have discussed,

$$T_j = T_j^{(1)} + T_j^{(2)} + T_j^{(g)}, \qquad (6.54)$$

gives the complete contribution of dimension-five operators in HQET to the structure functions. At this order in the operator product expansion only two matrix elements occur, λ_1 and λ_2. Furthermore, one of them, $\lambda_2 \simeq 0.12\ \mathrm{GeV}^2$, is known from $B^* - B$ mass splitting. The results for T_j determine the nonperturbative $\Lambda_{\rm QCD}^2/m_b^2$ corrections to the inclusive semileptonic decay rate.

6.3 Differential decay rates

The inclusive $\bar{B}$ semileptonic differential decay rate is calculated by using Eqs. (6.9) and (6.54), with the W_j's obtained from the imaginary part of the T_j's. The identity

$$-\frac{1}{\pi}\mathrm{Im}\left(\frac{1}{\Delta_0}\right)^{n+1} = \frac{(-1)^n}{n!}\delta^{(n)}\left[(m_b v - q)^2 - m_c^2\right], \tag{6.55}$$

where the superscript denotes the nth derivative of the δ function with respect to its argument, is useful in computing the W_j's. Terms with derivatives of the δ function are evaluated by first integrating by parts to take the derivatives off the δ function. In using this procedure, one must be careful to include the factor $\theta(4E_e E_{\nu_e} - q^2)$, which sets the lower limit of the E_{ν_e} integration, in the differential decay rate, since the derivative can act on this term. Differentiating the θ function with respect to E_{ν_e} gives

$$\delta\left[\left(\frac{m_b^2 - m_c^2 + q^2}{2m_b} - E_e\right) - \frac{q^2}{4E_e}\right], \tag{6.56}$$

which, in terms of the variables y, $\hat{q}^2$, and z defined in Eqs. (6.29) and (6.30), is the δ function $2\delta(z)/m_b$. This procedure gives for the differential decay rate

$$\begin{aligned}\frac{d\Gamma}{d\hat{q}^2\,dy} = \frac{G_F^2 m_b^5}{192\pi^3}|V_{cb}|^2 \Bigg\{ \theta(z)\Bigg[& 12(y - \hat{q}^2)(1 + \hat{q}^2 - \rho - y) \\ & - \frac{2\lambda_1}{m_b^2}(4\hat{q}^2 - 4\hat{q}^2\rho + 4\hat{q}^4 - 3y + 3\rho y - 6\hat{q}^2 y) \\ & - \frac{6\lambda_2}{m_b^2}(-2\hat{q}^2 - 10\hat{q}^2\rho + 10\hat{q}^4 - y + 5\rho y - 10\hat{q}^2 y)\Bigg] \\ & + \frac{\delta(z)}{y^2}\Bigg[-\frac{2\lambda_1}{m_b^2}(2\hat{q}^6 + \hat{q}^4 y^2 - 3\hat{q}^2 y^3 - \hat{q}^2 y^4 + y^5) \\ & - \frac{6\lambda_2}{m_b^2}\hat{q}^2(\hat{q}^2 - y)(5\hat{q}^2 - 8y + y^2)\Bigg] \\ & + \frac{\delta'(z)}{y^3}\left[-\frac{2\lambda_1}{m_b^2}\hat{q}^2(y^2 - \hat{q}^2)^2(y - \hat{q}^2)\right]\Bigg\},\end{aligned} \tag{6.57}$$

where the dimensionless variable $\hat{q}^2$, y, ρ, and z are defined in Eqs. (6.29) and (6.30). Experimentally, the electron energy spectrum $\mathrm{d}\Gamma/\mathrm{d}y$ is easier to study than the doubly differential decay rate. Integration of Eq. (6.57) over the allowed region $0 < \hat{q}^2 < y(1-y-\rho)/(1-y)$ gives

$$\frac{\mathrm{d}\Gamma}{\mathrm{d}y} = \frac{G_F^2 m_b^5}{192\pi^3}|V_{cb}|^2 \left\{ \left[2(3-2y)y^2 - 6y^2\rho - \frac{6y^2\rho^2}{(1-y)^2} + \frac{2(3-y)y^2\rho^3}{(1-y)^3} \right] \right.$$
$$- \frac{2\lambda_1}{m_b^2}\left[-\frac{5}{3}y^3 - \frac{y^3(5-2y)\rho^2}{(1-y)^4} + \frac{2y^3(10-5y+y^2)\rho^3}{3(1-y)^5} \right]$$
$$- \frac{6\lambda_2}{m_b^2}\left[-y^2\frac{(6+5y)}{3} + \frac{2y^2(3-2y)\rho}{(1-y)^2} \right.$$
$$\left.\left. + \frac{3y^2(2-y)\rho^2}{(1-y)^3} - \frac{5y^2(6-4y+y^2)\rho^3}{3(1-y)^4} \right] \right\}. \tag{6.58}$$

Integrating over the allowed electron energy $0 < y < 1-\rho$ yields the total $\bar{B} \to X_c e \bar{\nu}_e$ decay rate,

$$\Gamma = \frac{G_F^2 m_b^5}{192\pi^3}|V_{cb}|^2 \left[(1 - 8\rho + 8\rho^3 - \rho^4 - 12\rho^2 \ln\rho) \right.$$
$$+ \frac{\lambda_1}{2m_b^2}(1 - 8\rho + 8\rho^3 - \rho^4 - 12\rho^2 \ln\rho)$$
$$\left. - \frac{3\lambda_2}{2m_b^2}(3 - 8\rho + 24\rho^2 - 24\rho^3 + 5\rho^4 + 12\rho^2 \ln\rho) \right], \tag{6.59}$$

which can be written in the compact form

$$\Gamma = \frac{G_F^2 m_b^5}{192\pi^3}|V_{cb}|^2 \left[1 + \frac{\lambda_1}{2m_b^2} + \frac{3\lambda_2}{2m_b^2}\left(2\rho\frac{d}{d\rho} - 3 \right) \right] f(\rho), \tag{6.60}$$

where

$$f(\rho) = 1 - 8\rho + 8\rho^3 - \rho^4 - 12\rho^2 \ln\rho. \tag{6.61}$$

The first term is the leading term in the $m_b \to \infty$ limit and is equal to the free quark decay rate. The next two terms are $1/m_b^2$ corrections. The $1/m_b$ correction vanishes. Note that the ρ dependence of the coefficient of λ_1 is the same as that in the free quark decay rate. We will give a simple physical reason for this result in the next section.

Results for semileptonic $\bar{B}$-meson decays from the $b \to u$ transition are obtained from Eqs. (6.57), (6.58), and (6.59) by taking the limit $\rho \to 0$. Taking this limit is straightforward, except in the case of the electron spectrum in Eq. (6.58).

Suppose the electron energy spectrum in the $b \to c$ case contains a term of the form

$$g_\rho(y) = \frac{\rho^{n-1}}{(1-y)^n}. \tag{6.62}$$

The limit as $\rho \to 0$ of $g_\rho(y)$ is not zero. The problem is that the maximum value of y is $1-\rho$ and hence at maximum electron energy the denominator in Eq. (6.62) goes to zero as $\rho \to 0$. Imagine integrating $g_\rho(y)$ against a smooth test function $t(y)$. Integrating by parts

$$\begin{aligned}\lim_{\rho\to 0}\int_0^{1-\rho} dy\, t(y) g_\rho(y) &= \frac{1}{(n-1)}\left[t(1) - \lim_{\rho\to 0}\int_0^{1-\rho} dy\, \frac{dt}{dy}(y)\frac{\rho^{n-1}}{(1-y)^{n-1}}\right]\\ &= \frac{1}{(n-1)} t(1).\end{aligned} \tag{6.63}$$

Hence we conclude that

$$\lim_{\rho\to 0} g_\rho(y) = \frac{1}{(n-1)}\delta(1-y). \tag{6.64}$$

Differentiating the above gives

$$\lim_{\rho\to 0}\frac{\rho^{n-1}}{(1-y)^{n+1}} = -\frac{1}{n(n-1)}\delta'(1-y). \tag{6.65}$$

The $\rho \to 0$ limit of the electron spectrum in Eq. (6.58) is the $\bar{B} \to X_u e\bar{\nu}_e$ electron energy spectrum,

$$\begin{aligned}\frac{d\Gamma}{dy} = \frac{G_F^2 m_b^5 |V_{ub}|^2}{192\pi^3}\bigg\{&2(3-2y)y^2\theta(1-y)\\ &-\frac{2\lambda_1}{m_b^2}\left[-\frac{5}{3}y^3\theta(1-y) + \frac{1}{6}\delta(1-y) + \frac{1}{6}\delta'(1-y)\right]\\ &-\frac{2\lambda_2}{m_b^2}\left[-y^2(6+5y)\theta(1-y) + \frac{11}{2}\delta(1-y)\right]\bigg\},\end{aligned} \tag{6.66}$$

and the total decay width is

$$\Gamma = \frac{G_F^2 m_b^5 |V_{ub}|^2}{192\pi^3}\left(1 + \frac{\lambda_1}{2m_b^2} - \frac{9\lambda_2}{2m_b^2}\right). \tag{6.67}$$

6.4 Physical interpretation of $1/m_b^2$ corrections

The corrections to the decay rate proportional to λ_1 have a simple physical interpretation. They arise from the motion of the b quark inside the $\bar{B}$ meson. At leading order in the $1/m_b$ expansion, the b quark is at rest in the $\bar{B}$-meson rest frame, and the $\bar{B}$-meson differential decay rate is equal to the b-quark, decay rate, $\mathrm{d}\Gamma^{(0)}(v_r, m_b)$. However, in a $\bar{B}$ meson the b quark really has (in the $\bar{B}$-meson rest frame) a four momentum $p_b = m_b v_r + k$. We can consider this as a b quark with an effective mass m_b' and an effective four velocity v' satisfying

$$m_b' v' = m_b v_r + k. \tag{6.68}$$

Including effects of the b-quark motion in the $\bar{B}$ meson, we find the fully differential semileptonic decay rate $\mathrm{d}\Gamma$ is

$$\mathrm{d}\Gamma = \left\langle \mathrm{d}\Gamma^{(0)}(v', m_b')/v'^0 \right\rangle, \tag{6.69}$$

where v'^0 is the time-dilation factor, the fences denote averaging over k, and $\mathrm{d}\Gamma^{(0)}$ is the free b-quark differential decay rate. This averaging is done by expanding Eq. (6.69) to quadratic order in k and using

$$\langle k^\alpha \rangle = -\frac{\lambda_1}{2m_b} v_r^\alpha, \qquad \langle k^\alpha k^\beta \rangle = \frac{\lambda_1}{3}\left(g^{\alpha\beta} - v_r^\alpha v_r^\beta\right). \tag{6.70}$$

More powers of k would correspond to higher dimension operators in the OPE than those we have considered so far. In expanding Eq. (6.69) one can use

$$m_b'^2 = (m_b' v')^2 = (m_b v_r + k)^2 = m_b^2 + 2m_b v_r \cdot k + k^2. \tag{6.71}$$

Note that Eqs. (6.70) and (6.71) imply that $\langle m_b'^2 \rangle = \langle m_b^2 \rangle$. Since $v_{r\alpha}\langle k^\alpha k^\beta \rangle = 0$, we can replace m_b' by m_b in Eq. (6.69) without worrying about cross terms in the average where one factor of k arises from expanding m_b' and the other from expanding v'. The effective four velocity v' is related to v_r and k by

$$v'_\alpha = v_{r\alpha} + \frac{1}{m_b'} k_\alpha = v_{r\alpha} + \frac{k_\alpha}{m_b}, \tag{6.72}$$

so the time-dilation factor is

$$v'_0 = v_r \cdot v' = 1 + v_r \cdot k/m_b. \tag{6.73}$$

Averaging this yields $\langle v'_0 \rangle = 1 - \lambda_1/2m_b^2$, and since $v_{r\alpha}\langle k^\alpha k^\beta \rangle = 0$ we can replace the factor $1/v'^0$ in Eq. (6.69) by $(1 + \lambda_1/2m_b^2)$. The fully differential decay rate can be taken to be $\mathrm{d}\Gamma/\mathrm{d}\hat{q}^2\,\mathrm{d}y\,\mathrm{d}x$, where we have introduced the dimensionless neutrino energy variable

$$x = \frac{2E_{\nu_e}}{m_b}. \tag{6.74}$$

The variables x and y depend on the four velocity v_r of the b quark through $y = 2v_r \cdot p_e/m_b$, $x = 2v_r \cdot p_{\nu_e}/m_b$, and consequently, under the replacement

$v_r \to v'$,

$$y \to y' = y + \frac{2k \cdot p_e}{m_b^2}, \qquad x \to x' = x + \frac{2k \cdot p_{\nu_e}}{m_b^2}. \tag{6.75}$$

Hence Eq. (6.69) implies that

$$\frac{d\Gamma}{d\hat{q}^2\, dy\, dx} = \left[1 - \frac{\lambda_1}{2m_b^2}\left(-1 + y\frac{\partial}{\partial y} + x\frac{\partial}{\partial x} + \frac{1}{3}y^2\frac{\partial^2}{\partial y^2} + \frac{1}{3}x^2\frac{\partial^2}{\partial x^2} + \frac{2}{3}(xy - 2\hat{q}^2)\frac{\partial^2}{\partial x\, \partial y}\right)\right]\frac{d\Gamma^{(0)}}{d\hat{q}^2\, dy\, dx}. \tag{6.76}$$

Integrating over x yields

$$\frac{d\Gamma}{d\hat{q}^2\, dy} = \left[1 - \frac{\lambda_1}{2m_b^2}\left(-\frac{4}{3} + \frac{1}{3}y\frac{\partial}{\partial y} + \frac{1}{3}y^2\frac{\partial^2}{\partial y^2}\right)\right]\frac{d\Gamma^{(0)}}{d\hat{q}^2\, dy}, \tag{6.77}$$

where the free b-quark differential decay rate $d\Gamma^{(0)}/d\hat{q}^2\, dy$ is given in Eq. (6.28). Integrating over $\hat{q}^2$ and y, we find for the total decay rate

$$\Gamma = \left(1 + \frac{\lambda_1}{2m_b^2}\right)\Gamma^{(0)}. \tag{6.78}$$

Equations (6.77) and (6.78) give the correct λ_1 dependence of the $\bar{B}$-meson differential decay rates. Unfortunately, the dependence of the $\bar{B}$-meson differential decay rates in Eqs. (6.57)–(6.59) on λ_2 does not seem to have as simple a physical interpretation.

6.5 The electron endpoint region

The predictions that follow from the operator product expansion for the differential $\bar{B} \to X e \bar{\nu}_e$ semileptonic decay rate cannot be compared directly with experiment in all regions of the phase space. For example, the expression for the differential cross section $d\Gamma/d\hat{q}^2\, dy$ in Eq. (6.57) contains singular terms on the boundary of the Dalitz plot, $z = 0$. Rigorously, predictions based on the operator product expansion and perturbative QCD can be compared with experiment only when averaged over final hadronic state masses m_X with a smooth weighting function. Very near the boundary of the Dalitz plot, only the lower-mass final hadronic states can contribute, and the integration over neutrino energies does not provide the smearing over final hadronic masses needed to compare the operator product expansion results with experiment. In fact, since m_X is necessarily less than m_B, the weighting function is never truly smooth. As a result the contour integral over $v \cdot q$ needed to recover the structure functions W_j from those associated with the time-ordered product T_j necessarily pinches the cut at one point. Near the cut the use of the OPE cannot be rigorously justified because there will be propagators that have denominators close to zero. This is not considered a problem in a region where the final hadronic states are above the ground state by a large amount compared with the nonperturbative scale of the strong interactions, because threshold effects that are present in nature but not in the OPE analysis are very small. In inclusive $\bar{B}$ decay we assume that threshold effects associated with the limit on the maximum available hadronic mass $m_{X^{\max}}$ are negligible as long as $m_{X^{\max}} - m_{X^{\min}} \gg \Lambda_{\rm QCD}$. Near the boundary of the Dalitz plot this inequality is not satisfied. Note that at the order in α_s to which we have worked, the singularities in T_j are actually poles located at the ends of what we have called cuts. When α_s corrections are included, the singularities become the cuts we have described. Hence when radiative corrections are neglected, the contour in Fig. 6.3 need not be near a singularity in the $b \to c$ decay case. For $b \to u$ at q^2 near $q^2_{\max}$, the contour necessarily comes near singularities because the ends of the cuts are close together.

The endpoint region of the electron spectrum in inclusive semileptonic $\bar{B}$ decay has played an important role in determining the value of the element of the CKM matrix $|V_{ub}|$. For a given hadronic final state mass m_X, the maximum electron energy is $E_e^{\max} = (m_B^2 - m_X^2)/2m_B$. Consequently, electrons with energies greater than $E_e = (m_B^2 - m_D^2)/2m_B$ must necessarily come from the $b \to u$ transition. However, this endpoint region is precisely where the singular contributions proportional to $\delta(1-y)$ and $\delta'(1-y)$ occur in the $b \to u$ electron energy spectrum. Note that these singular terms occur at the endpoint set by quark–gluon kinematics, $E_e = m_b/2$, which is smaller than the true maximum $E_e^{\max} = m_B/2$. Clearly, in this region we must average over electron energies before comparing the predictions of the OPE and perturbative QCD with experiment.

To quantify the size of the averaging region in electron energies needed, we examine the general structure of the OPE. The most singular terms in the endpoint region result from expanding the k dependence in the denominator of the charm quark propagator. A term with power k^p produces an operator with p covariant derivatives and gives a factor of $1/\Delta_0^{p+1}$ in the T_j. This results in a factor of $\delta^{(p-1)}(1-y)$ in the electron energy spectrum. Matrix elements of operators with p covariant derivatives are of the order of $\Lambda_{\rm QCD}^p$ and so the general structure of the OPE prediction for the electron energy spectrum is

$$\begin{aligned}\frac{{\rm d}\Gamma}{{\rm d}y} \propto\ & \theta(1-y)(\varepsilon^0 + 0\,\varepsilon + \varepsilon^2 + \cdots)\\ & + \delta(1-y)(0\,\varepsilon + \varepsilon^2 + \cdots)\\ & + \delta'(1-y)(\varepsilon^2 + \varepsilon^3 + \cdots)\\ & \ \vdots\\ & + \delta^{(n)}(1-y)(\varepsilon^{n+1} + \varepsilon^{n+2} + \cdots)\\ & \ \vdots \end{aligned} \tag{6.79}$$

where ε^n denotes a quantity of the order of $(\Lambda_{\rm QCD}/m_b)^n$. It may contain smooth

y dependence. The zeroes are the coefficients of the dimension-four operators, which vanish by the equations of motion. Although the theoretical expression for $d\Gamma/dy$ is singular near the b-quark decay endpoint $y = 1$, the total semileptonic decay rate is not. The contribution to the total rate of a term of order $\varepsilon^m \delta^{(n)}(1-y)$ is order ε^m, and so the semileptonic width has a well-behaved expansion in powers of $1/m_b$:

$$\Gamma \propto (\varepsilon^0 + 0\,\varepsilon + \varepsilon^2 + \varepsilon^3 + \cdots). \tag{6.80}$$

In the endpoint region consider integrating $d\Gamma/dy$ against a normalized function of y that has a width σ. This provides a smearing of the electron energy spectrum near $y = 1$ and corresponds to examining the energy spectrum with resolution in y of σ (i.e., a resolution in electron energy of $m_b\sigma$). A meaningful prediction for the endpoint spectrum can be made when the smearing width σ is large enough that terms that have been neglected in Eq. (6.79) are small in comparison to the terms that have been retained. The singular term $\varepsilon^m \delta^{(n)}(1-y)$ (where $m > n$) smeared over a region of width σ gives a contribution of order $\varepsilon^m/\sigma^{n+1}$. If the smearing width σ is of the order of ε^p, the generic term $\varepsilon^m \delta^{(n)}(1-y)$ yields a contribution to the smeared spectrum of the order of $\varepsilon^{m-p(n+1)}$. Even though $m > n$, higher-order terms in the $1/m_b$ expansion get more important than lower-order ones unless $p \le 1$.

If the smearing in y is chosen of order ε (i.e., a region of electron energies of the order of Λ_{QCD}), then all terms of the form $\theta(1-y)$ and $\varepsilon^{n+1}\delta^{(n)}(1-y)$ contribute equally to the smeared electron energy spectrum, with less singular terms being suppressed. For example, all terms of order $\varepsilon^{n+2}\delta^{(n)}(1-y)$ are suppressed by ε, and so on. Thus one can predict the endpoint region of the electron energy spectrum, with a resolution in electron energies of the order of Λ_{QCD}, if these leading singular terms are summed. The sum of these leading singularities produces a contribution to $d\Gamma/dy$ of width ε but with a height of the same order as the free quark decay spectrum.

We can easily get the general form of the most singular contributions to the operator product expansion for the electron spectrum by using the physical picture of smearing over b-quark momenta discussed in the previous section. We want to continue the process to arbitrary orders in k, but only the most singular y dependence is needed. It arises only from the dependence of y on m_b and v. Shifting to m_b' and v',

$$y \to y' = \frac{2v' \cdot p_e}{m_b'} = y + k^\mu \frac{2}{m_b}(\hat{p}_{e\mu} - y v_\mu) + \cdots, \tag{6.81}$$

where the ellipsis denotes terms higher order in k, and $\hat{p}_e = p_e/m_b$. The term proportional to $y v_\mu$ in Eq. (6.81) arose from the dependence of m_b' on k. The most singular terms come from the y dependence in the factor $\theta(1-y)$, and so

$$\begin{aligned}\frac{d\Gamma}{dy} = \frac{d\Gamma^{(0)}}{dy}\Bigg[1 &+ \langle k^{\mu_1}\rangle \left(\frac{2}{m_b}\right)(\hat{p}_e - v)_{\mu_1}\frac{\partial}{\partial y} + \cdots \\ &+ \frac{1}{n!}\langle k^{\mu_1}\cdots k^{\mu_n}\rangle \left(\frac{2}{m_b}\right)^n (\hat{p}_e - v)_{\mu_1}\cdots(\hat{p}_e - v)_{\mu_n}\frac{\partial^n}{\partial y^n} + \cdots\Bigg]\theta(1-y).\end{aligned} \tag{6.82}$$

Equation (6.82) sums the most singular nonperturbative corrections in the endpoint region, provided one interprets the averaging over residual momenta as

$$\langle k_{\mu_1}\cdots k_{\mu_n}\rangle = \frac{1}{2}\langle \bar{B}(v)|\bar{b}_v iD_{(\mu_1}\cdots iD_{\mu_n)}b_v|\bar{B}(v)\rangle. \tag{6.83}$$

There is no operator ordering ambiguity because $\langle k^{\mu_1}\cdots k^{\mu_n}\rangle$ is contracted with a tensor completely symmetric in $\mu_1\cdots\mu_n$. Finally, only the part of the matrix element $\langle \bar{B}(v)|\bar{b}_v iD_{(\mu_1}\cdots iD_{\mu_n)}b_v|\bar{B}(v)\rangle$ proportional to $v_{\mu_1}\cdots v_{\mu_n}$ contributes to the most singular terms. A dependence on the metric tensor $g_{\mu_i\mu_j}$ would result in a factor of $(\hat{p}_e - v)^2$ that vanishes at $y=1$. So writing

$$\frac{1}{2}\langle \bar{B}(v)|\bar{b}_v iD_{(\mu_1}\cdots iD_{\mu_n)}b_v|\bar{B}(v)\rangle = A_n v_{\mu_1}\cdots v_{\mu_n} + \cdots, \tag{6.84}$$

we find the differential decay spectrum near $y=1$ is

$$\frac{d\Gamma}{dy} = \frac{d\Gamma^{(0)}}{dy}[\theta(1-y) + S(y)], \tag{6.85}$$

where the shape function $S(y)$ is

$$S(y) = \sum_{n=1}^{\infty}\frac{A_n}{m_b^n n!}\delta^{(n-1)}(1-y). \tag{6.86}$$

In Sec. 6.2.2, we showed that $A_1 = 0$ and $A_2 = -\frac{1}{3}\lambda_1$. At present, one must use phenomenological models for the shape function to extract $|V_{ub}|$ from semileptonic decay data in the endpoint region. This yields $|V_{ub}| \approx 0.1|V_{cb}|$. The perturbative QCD corrections to $d\Gamma/dy$ also become singular as $y \to 1$. These singular terms must also be summed to make a prediction for the shape of the electron spectrum in the endpoint region.

For inclusive $b \to c$ semileptonic decay, the $1/m_b^2$ corrections are not singular at the endpoint of the electron spectrum, but they are large because $m_c^2/m_b^2 \simeq 1/10$ is small. (At order $1/m_b^3$ singular terms occur even for $b \to c$ semileptonic decay.) It is instructive to plot the $b \to c$ electron spectrum including the $1/m_b^2$ corrections. This is done in Fig. 6.6. One can clearly see that the $1/m_b^2$ corrections become large near the endpoint. The OPE analysis gives the electron spectrum in Eq. (6.58), which depends on the heavy quark masses m_b and m_c. In particular,

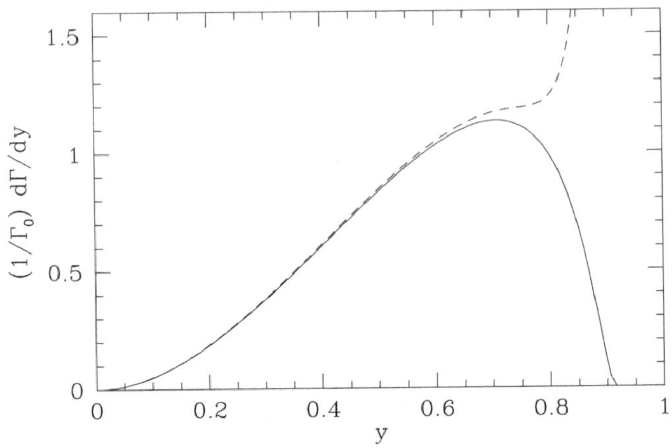

Fig. 6.6. The electron energy spectrum in inclusive semileptonic $\bar{B} \to X_c$ decay at lowest order (solid curve) and including the $1/m_b^2$ corrections (dashed curve), with $\lambda_1 = -0.2\ \text{GeV}^2$, $m_b = 4.8$ GeV, and $m_c = 1.4$ GeV. Here $\Gamma_0 = G_F^2 |V_{cb}|^2 m_b^5 / 192\pi^3$.

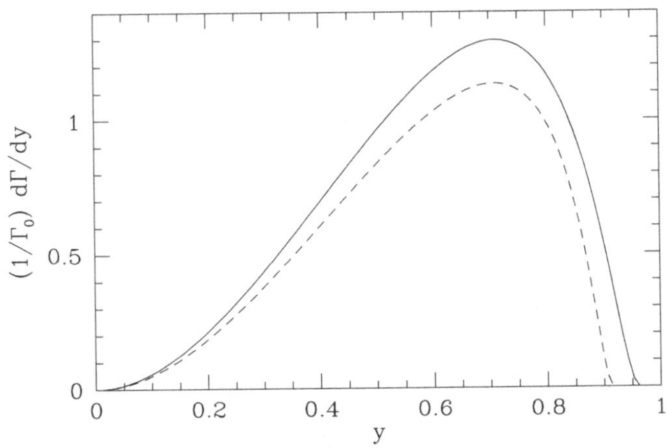

Fig. 6.7. The electron energy spectrum in inclusive semileptonic $\bar{B} \to X_c$ decay, using the lowest-order formula with quark masses (dashed curve) and with hadron masses (solid curve); y is defined as $2E_e/m_b$ in both plots, and $\Gamma_0 = G_F^2 |V_{cb}|^2 m_b^5 / 192\pi^3$.

the electron endpoint energy is $(m_b^2 - m_c^2)/2m_b$. The true kinematic endpoint for the electron spectrum is $(m_B^2 - m_D^2)/2m_B$, and it depends on the hadron masses. In Fig. 6.7, the lowest-order electron spectrum using quark masses has been compared with the same spectrum in which quark masses have been replaced by hadron masses. Over most of the phase space, this is close to the true spectrum, but very near the maximum value of E_e there is no theoretical basis to believe that the lowest-order spectrum with hadron masses has any connection with the actual electron spectrum. Nevertheless, the spectrum with hadron masses ends

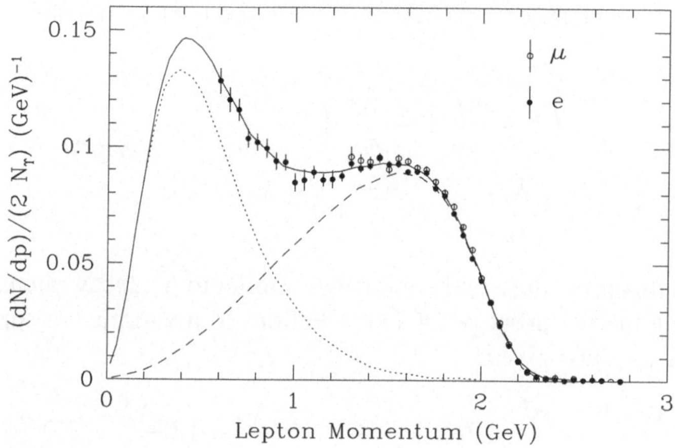

Fig. 6.8. The inclusive lepton energy spectrum for semileptonic $\bar{B} \to X_c$ decay, as measured by the CLEO Collaboration. The data are from the Ph.D thesis of R. Wang. The filled dots are the electron spectrum, and the open dots are the muon spectrum. The dashed curve is a model fit to the primary leptons from $b \to c$ semileptonic decay, which should be compared with theoretical predictions in Figs. 6.6 and 6.7. The dotted curve is a model fit to the secondary leptons from semileptonic decay of the c quark produced in b decay, and the solid curve is the sum of the two.

at the true kinematic endpoint of the allowed electron spectrum. The measured inclusive lepton spectrum in semileptonic B decay is shown in Fig. 6.8.

6.6 $|V_{cb}|$ from inclusive decays

The expression for the inclusive differential semileptonic decay rate in Eq. (6.57) can be used to deduce the HQET parameters $\bar{\Lambda}$ and λ_1. In addition, it provides a determination of the CKM matrix element V_{cb}. For comparison with experiment, it is useful to eliminate the c- and b-quark masses in favor of hadron masses. The average D- and B-meson masses are

$$\bar{m}_D = \frac{m_D + 3m_{D^*}}{4} = 1.975 \text{ GeV}, \qquad \bar{m}_B = \frac{m_B + 3m_{B^*}}{4} = 5.313 \text{ GeV}. \tag{6.87}$$

Using the results of Chapter 4, we find

$$\begin{aligned} m_c &= \bar{m}_D - \bar{\Lambda} + \frac{\lambda_1}{2\bar{m}_D} + \cdots, \\ m_b &= \bar{m}_B - \bar{\Lambda} + \frac{\lambda_1}{2\bar{m}_B} + \cdots, \end{aligned} \tag{6.88}$$

where the ellipses denote terms higher order in the $1/m_Q$ expansion. This gives, for example,

$$\begin{aligned}\frac{m_c}{m_b} &= \frac{\bar{m}_D}{\bar{m}_B} - \frac{\bar{\Lambda}}{\bar{m}_B}\left(1 - \frac{\bar{m}_D}{\bar{m}_B}\right) - \frac{\bar{\Lambda}^2}{\bar{m}_B^2}\left(1 - \frac{\bar{m}_D}{\bar{m}_B}\right) + \frac{\lambda_1}{2\bar{m}_B\bar{m}_D}\left(1 - \frac{\bar{m}_D^2}{\bar{m}_B^2}\right)\\ &\simeq 0.372 - 0.63\frac{\bar{\Lambda}}{\bar{m}_B} - 0.63\frac{\bar{\Lambda}^2}{\bar{m}_B^2} + 1.2\frac{\lambda_1}{\bar{m}_B^2}.\end{aligned} \tag{6.89}$$

Applying this procedure to the inclusive semileptonic decay rate in Eq. (6.59) and including the perturbative QCD corrections to the terms not suppressed by powers of $\Lambda_{\rm QCD}/m_Q$ gives

$$\Gamma_{\rm SL}(B) = \frac{G_F^2|V_{cb}|^2 m_B^5}{192\pi^3}\, 0.369\left[\eta_\Gamma - 1.65\frac{\bar{\Lambda}}{\bar{m}_B} - 1.0\frac{\bar{\Lambda}^2}{\bar{m}_B^2} - 3.2\frac{\lambda_1}{\bar{m}_B^2}\right]. \tag{6.90}$$

Note that m_B^5 has been factored out instead of $\bar{m}_B^5$. This choice makes the coefficient of $\lambda_2/\bar{m}_B^2$ very small, and it has been neglected in the square brackets in Eq. (6.90).

The perturbative corrections to the leading term in the $1/m_Q$ expansion are known to order α_s^2:

$$\eta_\Gamma = 1 - 1.54\frac{\alpha_s(m_b)}{\pi} - 12.9\left[\frac{\alpha_s(m_b)}{\pi}\right]^2 = 0.83. \tag{6.91}$$

Using Eq. (6.90), the measured semileptonic branching ratio $\mathrm{BR}(B \to X e\bar{\nu}_e) = (10.41 \pm 0.29)\,\%$, and the B lifetime $\tau(B) = (1.60 \pm 0.04) \times 10^{-12}$ s, one finds

$$|V_{cb}| = \frac{[\,39 \pm 1\,(\text{exp})\,] \times 10^{-3}}{\sqrt{1 - 2.0\frac{\bar{\Lambda}}{\bar{m}_B} - 1.2\left(\frac{\bar{\Lambda}}{\bar{m}_B}\right)^2 - 3.9\frac{\lambda_1}{\bar{m}_B^2}}}. \tag{6.92}$$

The differential decay rate constrains the values of $\bar{\Lambda}$ and λ_1. An analysis of the electron energy spectrum gives (at order α_s^2) $\bar{\Lambda} \simeq 0.4$ GeV and $\lambda_1 \simeq -0.2$ GeV^2, with a large uncertainty. These values imply that $|V_{cb}| = 0.042$. Note that this is close to the value extracted from semileptonic $\bar{B} \to D^* e\bar{\nu}_e$ decay in Chapter 4 (see Eq. (4.65)). Theoretical uncertainty in this determination of V_{cb} arises from the values of $\bar{\Lambda}$ and λ_1 and possible violations of quark hadron duality.

In Eq. (6.91) the order α_s^2 term is ~60% of the order α_s term. There are two reasons for this. First, recall from Chapter 4 that $\bar{\Lambda}$ is not a physical quantity and has a renormalon ambiguity of the order of $\Lambda_{\rm QCD}$. Using HQET, we can relate $\bar{\Lambda}$ to a measurable quantity, for example $\langle \delta s_H \rangle$, the average value of $\delta s_H = s_H - \bar{m}_D^2$, where s_H is the hadronic invariant mass squared in semileptonic $\bar{B}$ decay. This relation involves a perturbative series in α_s. If one eliminates $\bar{\Lambda}$ in Eq. (6.90) in favor of $\langle \delta s_H \rangle$, then the combination of the perturbative series in the relation between $\bar{\Lambda}$ and $\langle \delta s_H \rangle$ and the series η_Γ will replace η_Γ in Eq. (6.90). This modified series has no Borel singularity at $u = 1/2$ and is somewhat better behaved. Second, the typical energy of the decay products in $b \to ce\bar{\nu}_e$ quark decay is not m_b, but rather $E_{\rm typ} \sim (m_b - m_c)/3 \sim 1.2$ GeV. Using this scale instead of m_b to evaluate the strong coupling at in Eq. (6.91) leads to a series in which the order α_s^2 term is 25% of the order α_s term. Note that for this one uses $\alpha_s(m_b) = \alpha_s(E_{\rm typ}) - \alpha_s^2(E_{\rm typ})\beta_0 \ln m_b^2/E_{\rm typ}^2 + \cdots$ in Eq. (6.91) and expands η_Γ to quadratic order in $\alpha_s(E_{\rm typ})$.

6.7 Sum rules

One can derive a set of sum rules that restrict exclusive $\bar{B} \to D^{(*)} e\bar{\nu}_e$ form factors by comparing the inclusive and exclusive semileptonic $\bar{B}$ decay rates. The basic ingredient is the simple result that the inclusive $\bar{B}$ decay rate must always be greater than or equal to the exclusive $\bar{B} \to D^{(*)}$ decay rate.

The analysis uses $T_{\alpha\beta}$ considered as a function of q_0 with $\mathbf{q}$ held fixed. It is convenient not to focus on just the left-handed current, which is relevant for semileptonic decay, but rather to allow J to be the axial vector or vector currents or a linear combination of these. Also we change variables from q_0 to

$$\varepsilon = m_B - q_0 - E_{X_c^{\rm min}}, \tag{6.93}$$

where $E_{X_c^{\rm min}} = \sqrt{m^2_{X_c^{\rm min}} + |\mathbf{q}|^2}$ is the minimal possible energy of the hadronic state. With this definition, $T_{\alpha\beta}(\varepsilon)$ has a cut in the complex ε plane along $0 < \varepsilon < \infty$, corresponding to physical states with a c quark. $T_{\mu\nu}$ has another cut for $2m_B - E_{X_{\bar{c}bb}^{\rm min}} - E_{X_c^{\rm min}} > \varepsilon > -\infty$ corresponding to physical states with two b quarks and a $\bar{c}$ quark.* This cut will not be important for the results in this section. Contracting $T_{\mu\nu}$ with a fixed four vector a^ν yields

$$a^{*\mu} T_{\mu\nu}(\varepsilon) a^\nu = -\sum_{X_c} (2\pi)^3 \delta^3(\mathbf{q} + \mathbf{p}_X) \frac{\langle \bar{B}|J^\dagger \cdot a^*|X_c\rangle\langle X_c|J \cdot a|\bar{B}\rangle}{2m_B \left(E_{X_c} - E_{X_c^{\rm min}} - \varepsilon\right)} + \cdots, \tag{6.94}$$

where the ellipsis denotes the contribution from the cut corresponding to two b quarks and a $\bar{c}$ quark. Consider integrating the product of a weight function $W_\Delta(\varepsilon)$ and $T_{\mu\nu}(\varepsilon)$ along the contour C shown in Fig. 6.9. Assuming W_Δ is analytic in the region enclosed by this contour, we get

$$\begin{aligned} &\frac{1}{2\pi i} \int_C d\varepsilon\, W_\Delta(\varepsilon)\, a^{*\mu} T_{\mu\nu}(\varepsilon) a^\nu \\ &\quad = \sum_{X_c} W_\Delta\big(E_{X_c} - E_{X_c^{\rm min}}\big)(2\pi)^3 \delta^3\big(\mathbf{q} + \mathbf{p}_X\big) \frac{|\langle X_c|J \cdot a|\bar{B}\rangle|^2}{2m_B}. \end{aligned} \tag{6.95}$$

* Note that the left-hand and right-hand cuts are exchanged when switching from q^0 to ε because of the minus sign in Eq. (6.93).

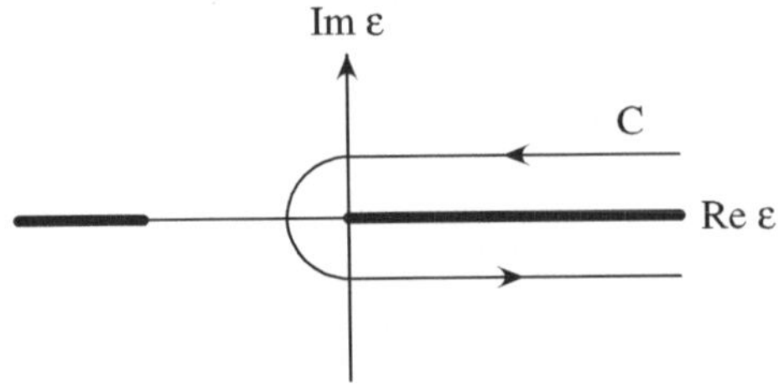

Fig. 6.9. The sum-rule cut.

We want the weight function $W_\Delta(\varepsilon)$ to be positive semidefinite along the cut so the contribution of each term in the sum over X_c above is nonnegative. For convenience we impose the normalization condition $W_\Delta(0) = 1$. We also assume W_Δ is flat near $\varepsilon = 0$ and falls off rapidly to zero for $\varepsilon \gg \Delta$. If the operator product expansion and perturbative QCD are used to evaluate the left-hand side of Eq. (6.95), it is crucial that W_Δ is flat in a region of ϵ much bigger than $\Lambda_{\rm QCD}$. Otherwise, higher-order terms in the operator product expansion and perturbative corrections will be large.

The positivity of each term in the sum over states X in Eq. (6.95) implies the bound

$$\frac{1}{2\pi i} \int_C d\varepsilon\, W_\Delta(\varepsilon)\, a^{*\mu} T_{\mu\nu}(\varepsilon) a^\nu > \frac{\left|\langle X_c^{\rm min}| J \cdot a |\bar{B}\rangle\right|^2}{4 m_B E_{X_c}^{\rm min}}. \tag{6.96}$$

To derive this we note that the sum over X_c includes an integral over $d^3 p/(2\pi)^3 2E$ for each particle in the final state. For the one-particle state, $X_c^{\rm min}$, performing the integral over its three momentum by using the delta function leaves the factor $(2E_{X_c^{\rm min}})$ in the denominator of Eq. (6.96). All the other states make a nonnegative contribution, leading to the inequality Eq. (6.96).

A set of possible weight functions is

$$W_\Delta^{(n)} = \frac{\Delta^{2n}}{\varepsilon^{2n} + \Delta^{2n}}. \tag{6.97}$$

For $n > 2$ the integral Eq. (6.96) is dominated by states with a mass less than Δ. These weight functions have poles at $\varepsilon = (-1)^{1/2n}\Delta$. Therefore, if n is not too large and Δ is much greater than the QCD scale, the contour in Fig. 6.9 is far from the cut. As $n \to \infty$, $W_\Delta^{(n)} \to \theta(\Delta - \varepsilon)$ for positive ε, which corresponds to summing over all final hadronic resonances with equal weight up to excitation energy Δ. In this case the poles of W_Δ approach the cut and the contour in Fig. 6.9 must be deformed to touch the cut at $\varepsilon = \Delta$. As in the semileptonic decay rate, this is usually not considered a problem as long as $\Delta \gg \Lambda_{\rm QCD}$. Here $W_\Delta^{(\infty)}$ is the common choice for the weight function, and we use it for the remainder of this chapter.

To illustrate the utility of Eq. (6.96) we go over to HQET, where the charm and bottom quark masses are taken as infinite, and let $J^\mu = \bar{c}_{v'} \gamma^\mu b_v$ and $a^\mu = v^\mu$.

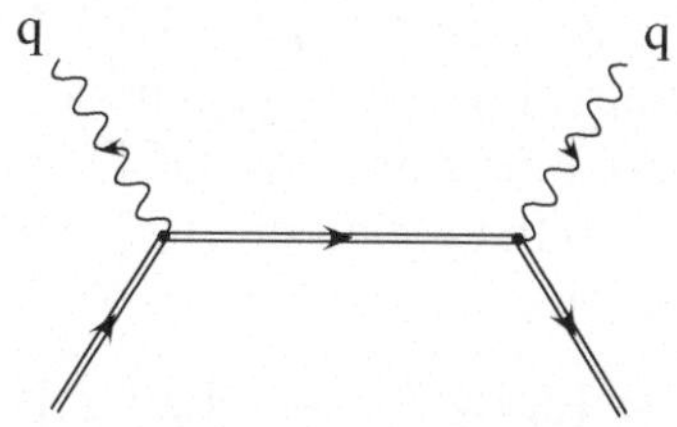

Fig. 6.10. The leading-order diagram for the OPE.

Only the pseudoscalar member of the ground state D, D^* doublet contributes in this case, and

$$\langle D(v')|J \cdot v|\bar{B}(v)\rangle = (1 + w)\xi(\omega), \tag{6.98}$$

where $w = v \cdot v'$. For Δ large compared with Λ_{QCD}, the leading contribution to the time-ordered product $T_{\mu\nu}(\varepsilon)$ comes from performing the OPE, evaluating the coefficients to lowest order in α_s, and keeping only the lowest-dimension operators. We work in the $\bar{B}$-meson rest frame $v = v_r$ and define the four velocity of the charm quark by $-\mathbf{q} = m_c\mathbf{v}'$. Then the charm quark's residual momentum is $(k^0 = m_b v_r^0 - q^0 - m_c v'^0, \mathbf{k} = 0)$. In this frame $v_0' = v_r \cdot v' = w$. The leading operator in the OPE is $\bar{b}_{v_r} b_{v_r}$, and its coefficient follows from the Feynman diagram for the b-quark matrix element shown in Fig. 6.10. This yields

$$v_r^\mu T_{\mu\nu}(\varepsilon) v_r^\nu = \frac{(v_r \cdot v' + 1)}{2v_0'(m_b v_{r0} - q_0 - m_c v_0')}. \tag{6.99}$$

The variable ε defined in Eq. (6.93) can be expressed in terms of the heavy quark masses, $\bar{\Lambda}$ and w,

$$\begin{aligned} \varepsilon &= m_b + \bar{\Lambda} - q_0 - \sqrt{(m_c + \bar{\Lambda})^2 + m_c^2(w^2 - 1)} + \cdots \\ &= m_b - q_0 - m_c w + \frac{\bar{\Lambda}\,(w - 1)}{w} + \cdots, \end{aligned} \tag{6.100}$$

where the ellipses denote terms suppressed by powers of $\Lambda_{\text{QCD}}/m_{b,c}$. Using this, we find Eq. (6.99) becomes

$$v_r^\mu T_{\mu\nu}(\varepsilon) v_r^\nu = \left(\frac{w + 1}{2w}\right) \frac{1}{\varepsilon - \bar{\Lambda}(w - 1)/w}. \tag{6.101}$$

Performing the contour integration gives

$$\frac{w + 1}{2w} > \frac{|\xi(w)|^2 (1 + w)^2}{4w}. \tag{6.102}$$

At zero recoil, $\xi(1) = 1$, and the above bound is saturated. Writing $\rho^2 = -\mathrm{d}\xi/\mathrm{d}w|_{w=1}$, we find the above gives the Bjorken bound on the slope of the Isgur-Wise function at zero recoil, $\rho^2 \geq 1/4$. Away from zero recoil the Isgur-Wise function is subtraction-point dependent and consequently ρ^2 depends on the subtraction point. Perturbative QCD corrections add terms of the form $\alpha_s(\mu)(\ln \Delta^2/\mu^2 + C)$ to the bound on ρ^2. Consequently the bound on ρ^2 is more correctly written as

$$\rho^2(\Delta) \geq 1/4 + \mathcal{O}[\alpha_s(\Delta)]. \tag{6.103}$$

6.8 Inclusive nonleptonic decays

The nonleptonic weak decay Hamiltonian for $b \to c\bar{u}d$ decays $H_W^{(\Delta c=1)}$ was given in Eqs. (1.124) and (1.125). The nonleptonic decay rate is related to the imaginary part of the B-meson matrix element of the time-ordered product of this Hamiltonian with its Hermitian conjugate,

$$t = i \int d^4x\, T\big[H_W^{(\Delta c=1)\dagger}(x)\, H_W^{(\Delta c=1)}(0)\big]. \tag{6.104}$$

Taking the matrix element of t between B-meson states at rest and inserting a complete set of states between the two Hamiltonian densities yields

$$\begin{aligned}\Gamma^{(\Delta c=1)} &= \sum_X (2\pi)^4\, \delta^4\,(p_B - p_X)\,\frac{\left|\langle X(p_X)|H_W^{(\Delta c=1)}(0)|\bar{B}(p_B)\rangle\right|^2}{2m_B} \\ &= \frac{\operatorname{Im}\langle \bar{B}|t|\bar{B}\rangle}{m_B},\end{aligned} \tag{6.105}$$

where the first line is the definition of $\Gamma^{(\Delta c=1)}$.

Inclusive nonleptonic decays can also be studied by using the OPE. In the case of semileptonic decays, one can smear the decay distributions over the leptonic kinematic variables q^2 and $q \cdot v$. The corresponding smearing variables do not exist for nonleptonic decay, since all the final-state particles are hadrons. For nonleptonic decays, one needs the additional assumption that the OPE answer is correct even without averaging over the hadron invariant mass, which is fixed to be the B-meson mass. This assumption is reasonable because m_B is much greater than $\Lambda_{\rm QCD}$. The leading term in the OPE is computed from the diagram in Fig. 6.11. Its imaginary part gives the total nonleptonic decay width. The

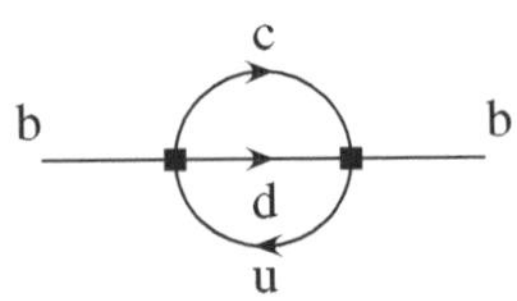

Fig. 6.11. OPE diagram for inclusive nonleptonic decay.

situation in the case of nonleptonic decays is not very different from the case of semileptonic decays, since there the contour of $v \cdot q$ integration cannot be deformed so that it is always far from the physical cut; see Fig. 6.3.

One can compare the OPE computation of the nonleptonic decay width with that of the semileptonic decay distribution. Imagine evaluating Fig. 6.11 with $\bar{u}d$ replaced by $\bar{\nu}e$. Computing the imaginary part of the diagram is equivalent to evaluating the phase space integral for the final-state fermions. Thus performing an OPE from the imaginary part of Fig. 6.11 is equivalent to integrating the decay distributions to obtain the total decay width in Eq. 6.60. In the case of nonleptonic decays, only the total width can be computed. Decay distributions are not accessible using this method. Another difference between the semileptonic and nonleptonic decays is that the weak Hamiltonian $H_W^{(\Delta c=1)}$ contains two terms with coefficients $C_1(m_b)$ and $C_2(m_b)$ due to summing radiative corrections, using the renormalization group equations.

Including $\Lambda_{\rm QCD}^2/m_b^2$ terms, we find the final result for the nonleptonic decay width $\Gamma^{(\Delta c=1)}$ computed using the OPE together with a transition to HQET is

$$\Gamma^{(\Delta c=1)} = 3\frac{G_F^2 m_b^5}{192\pi^3}|V_{cb}V_{ud}|^2 \left\{ \left(C_1^2 + \frac{2}{3}C_1C_2 + C_2^2\right)\left[\left(1 + \frac{\lambda_1}{2m_b^2}\right) + \frac{3\lambda_2}{2m_b^2}\left(2\rho\frac{d}{d\rho} - 3\right)\right] f(\rho) - 16C_1C_2\frac{\lambda_2}{m_b^2}(1-\rho)^3 \right\}, \qquad (6.106)$$

where $f(\rho)$ was defined in Eq. (6.61) and $C_{1,2}$ are evaluated at $\mu = m_b$.

The form of the leading-order term was computed in Problem 8 in Chapter 1. The order $\Lambda_{\rm QCD}^2/m_b^2$ part of Eq. (6.106) proportional to λ_1 can be deduced using the techniques of Sec. (6.4). Equation (6.78) holds for both the semileptonic and nonleptonic decay widths. However, the correction proportional to λ_2 cannot be deduced as simply. Like the semileptonic decay case, it arises from two sources. One is from a b-quark matrix element of the time-ordered product t, where the b quarks have momentum $p_b = m_b v + k$. Expanding in the residual momentum k gives, at quadratic order in k, dependence on λ_2 through the transition from full QCD to HQET. This part of the λ_2 dependence is the same for the nonleptonic and semileptonic decays. There is also λ_2 dependence that is identified from the $b \to b +$ gluon matrix element. It is different in the nonleptonic decay case because of the possibility that the gluon is emitted off the d or $\bar{u}$ quarks, as shown in Fig. 6.12. This contribution depends on the color structure of the operators O_1 and O_2, and we consider the pieces in $\Gamma^{(\Delta c=1)}$ proportional to C_1^2, C_2^2, and C_1C_2 successively.

For the piece of the λ_2 term proportional to C_1^2, the contribution where a gluon attaches to a d or $\bar{u}$ quark vanishes by color conservation because these diagrams are proportional to ${\rm Tr}\, T^A = 0$. Consequently, the λ_2 dependence proportional to C_1^2 is the same for nonleptonic and semileptonic decays. For the contribution

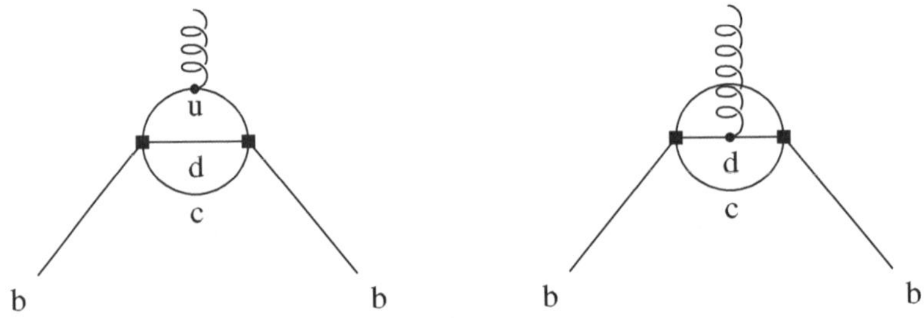

Fig. 6.12. OPE diagram for inclusive nonleptonic B decay with a gluon emitted from one of the light quark lines.

proportional to C_2^2, the $b \to b +$ gluon matrix element is the same as semileptonic $b \to u$ decay, provided the electron is not massless but rather has a mass equal to that of the c quark. This is easily seen after making a Fierz rearrangement of the quark fields in O_2. Note that the c-quark mass only enters the calculation of $T_{\mu\nu}$ through Δ_0. The left-handed projectors P_L remove the c-quark mass term in the numerator of its propagator. After taking the imaginary part, the m_c dependence in Δ_0 goes into setting the correct three-body phase space. However, the phase space is the same for $b \to c$ decay with massless leptons and $b \to u$ decay with a massless neutrino and the electron having the same mass as the c quark. Consequently, the $\lambda_2 C_2^2$ term is also the same as in semileptonic decay. For the $\lambda_2 C_1 C_2$ term there is the usual part that is the same as for semileptonic decays, as well as an additional contribution from the piece of the $b \to b +$ gluon matrix element of t where the gluon attaches to either the d or $\bar{u}$ quarks. This additional part is the last term in Eq. (6.106), and the remainder of this section is devoted to computing it.

The part of $\text{Im}\langle bg|t|b\rangle$ coming from Fig. 6.12 is

$$\begin{aligned} &16\pi i G_F^2 |V_{cb}|^2 |V_{ud}|^2 \int \frac{d^4q}{(2\pi)^4} \delta\left[(m_b v - q)^2 - m_c^2\right] \\ &\times \bar{u}\gamma^\mu P_L (m_b \not{v} - \not{q} + m_c)\gamma^\nu P_L u \,\text{Im}\,\Pi_{\mu\nu}. \end{aligned} \tag{6.107}$$

In Eq. (6.107) the δ function comes from the imaginary part of the c-quark propagator and

$$\begin{aligned} \Pi_{\mu\nu} = {} & g T^A \varepsilon^{A\lambda *} \int \frac{d^4k}{(2\pi)^4} \\ & \times \text{Tr}\Bigg[\gamma^\mu P_L \frac{\not{k} - \not{p}/2}{(k - p/2)^2 + i\varepsilon}\gamma_\lambda \frac{\not{k} + \not{p}/2}{(k + p/2)^2 + i\varepsilon}\gamma_\nu P_L \frac{\not{k} - \not{q}}{(k - q)^2 + i\varepsilon} \\ & + \gamma_\mu P_L \frac{\not{k} + \not{q}}{(k + q)^2 + i\varepsilon}\gamma_\nu P_L \frac{\not{k} - \not{p}/2}{(k - p/2)^2 + i\varepsilon}\gamma_\lambda \frac{\not{k} + \not{p}/2}{(k + p/2)^2 + i\varepsilon}\Bigg]. \end{aligned} \tag{6.108}$$

As in the semileptonic decay case, the gluon has outgoing momentum p and the initial and final b quarks have residual momentum $p/2$ and $-p/2$, respectively. Expanding in p, keeping only the linear term, combining denominators by using the Feynman trick, performing the k integration, taking the imaginary part, and performing the Feynman parameter integration gives

$$\text{Im}\,\Pi_{\mu\nu} = \frac{igp^\beta T^A \varepsilon^{A\lambda *}}{32\pi}\delta(q^2)\text{Tr}[\gamma_\mu(\gamma_\beta\gamma_\lambda \not{q} - \not{q}\gamma_\lambda\gamma_\beta)\gamma_\nu \not{q} P_R + (\mu \leftrightarrow \nu)]. \tag{6.109}$$

Only the part antisymmetric in β and λ gives a contribution of the type we are interested in. Performing the trace yields

$$\text{Im}\,\Pi_{\mu\nu} = \frac{gp^\beta T^A \varepsilon^{A\lambda *}}{4\pi}\delta(q^2)[\epsilon_{\beta\nu\lambda\alpha} q^\alpha q_\mu + (\mu \leftrightarrow \nu)]. \tag{6.110}$$

Putting this into Eq. (6.107), identifying the spinors with HQET b-quark fields, and using $p^\beta T^A \epsilon^{A\lambda *} \to -iG^{\beta\lambda}/2$ and Eq. (6.50) for the resulting B-meson matrix element, Fig. 6.12 gives the following contribution to the nonleptonic width:

$$\begin{aligned}\delta\Gamma^{(\Delta c=1)} &= -32 C_1 C_2 |V_{cb}|^2 |V_{ud}|^2 G_F^2 \lambda_2 \\ &\times \int \frac{d^4 q}{(2\pi)^4}\delta\left[(m_b v - q)^2 - m_c^2\right]\delta(q^2) m_b (v\cdot q)^2.\end{aligned} \tag{6.111}$$

Performing the q^0 and $\boldsymbol{q}$ integrations with the δ functions yields

$$\delta\Gamma^{(\Delta c=1)} = -\frac{C_1 C_2 |V_{cb}|^2 |V_{ud}|^2 G_F^2 \lambda_2 m_b^3}{4\pi^3}\left(1 - \frac{m_c^2}{m_b^2}\right)^3, \tag{6.112}$$

which is the last term in Eq. (6.106). The contribution of dimension-six four-quark operators to the nonleptonic width is thought to be more important than the dimension-five operators considered in this section, because their coefficients are enhanced by a factor of $16\pi^2$. The influence of similar four-quark operators in the case of $B_s - \bar{B}_s$ mixing will be considered in the next section.

6.9 $B_s - \bar{B}_s$ mixing

The light antiquark in a $\bar{B}$ or $\bar{B}_s$ meson is usually called the spectator quark, because at leading order in the OPE, its field does not occur in the operators whose matrix elements give the inclusive decay rate. This persists at order $1/m_b^2$ since $\lambda_{1,2}$ are defined as the matrix elements of operators constructed from b-quark and gluon fields. At order $1/m_b^3$, the spectator quark fields first appear because dimension-six four-quark operators of the form $\bar{b}_v b_v \bar{q} q$ occur in the OPE. These operators play a very important role in $B_s - \bar{B}_s$ width mixing.

Recall that $CP|B_s\rangle = -|\bar{B}_s\rangle$, so the CP eigenstates are

$$|B_{s1}\rangle = \frac{1}{\sqrt{2}}(|B_s\rangle + |\bar{B}_s\rangle)$$
$$|B_{s2}\rangle = \frac{1}{\sqrt{2}}(|B_s\rangle - |\bar{B}_s\rangle), \tag{6.113}$$

with $CP|B_{sj}\rangle = (-1)^j|B_{sj}\rangle$. At second order in the weak interactions there are $|\Delta b| = 2$, $|\Delta s| = 2$ processes that cause mass and width mixing between the $|B_s\rangle$ and $|\bar{B}_s\rangle$ states. In the limit that CP is conserved, it is the states $|B_{sj}\rangle$ rather than $|B_s\rangle$ and $|\bar{B}_s\rangle$ that are eigenstates of the effective Hamiltonian $H_{\rm eff} = \mathbb{M} + i\mathbb{W}/2$, where $\mathbb{M}$ and $\mathbb{W}$ are the 2×2 mass and width matrices for this system. For simplicity, we will neglect CP violation in the remainder of this section; it is straightforward to extend the arguments to include CP violation. In the $B_s - \bar{B}_s$ basis, the width matrix $\mathbb{W}$ is

$$\mathbb{W} = \begin{pmatrix} \Gamma_{B_s} & \Delta\Gamma \\ \Delta\Gamma & \Gamma_{\bar{B}_s} \end{pmatrix}. \tag{6.114}$$

CPT invariance implies that $\Gamma_{B_s} = \Gamma_{\bar{B}_s}$, so the widths of the eigenstates of $H_{\rm eff}$ are

$$\Gamma_j = \Gamma_{B_s} - (-1)^j\Delta\Gamma. \tag{6.115}$$

The difference between the widths of the two eigenstates $|B_{s1}\rangle$ and $|B_{s2}\rangle$ is $\Gamma_1 - \Gamma_2 = 2\Delta\Gamma$.

The width mixing element $\Delta\Gamma$ in Eq. (6.114) is defined by

$$\Delta\Gamma \equiv \sum_X (2\pi)^4\delta^4(p_B - p_X)\frac{\langle B_s|H_W^{(\Delta c=0)}|X\rangle\langle X|H_W^{(\Delta c=0)}|\bar{B}_s\rangle}{2m_{B_s}}$$
$$= \text{Im}\,\frac{\langle B_s|i\int d^4x\, T\big[H_W^{(\Delta c=0)}(x)\, H_W^{(\Delta c=0)}(0)\big]|\bar{B}_s\rangle}{2m_{B_s}}. \tag{6.116}$$

The first line is the definition of $\Delta\Gamma$, and the second line can be verified by inserting a complete set of states. There is a difference of a factor of 2 when compared with Eq. (6.105), because now both time orderings contribute. The width transition matrix element $\Delta\Gamma$ comes from final states that are common in B_s and $\bar{B}_s$ decay. For this reason it involves only the $\Delta c = 0$ part of the weak Hamiltonian; the $\Delta c = 1$ part does not contribute. The $\Delta c = 0$ part of the weak Hamiltonian gives at tree level the quark decay $b \to c\bar{c}s$. In the leading logarithmic approximation,

$$H_W^{(\Delta c=0)} = \frac{4G_F}{\sqrt{2}} V_{cb} V_{cs}^* \sum_i C_i(\mu) Q_i(\mu), \tag{6.117}$$

where the operators $Q_i(\mu)$ that occur are

$$\begin{aligned}
Q_1 &= (\bar{c}^\alpha \gamma_\mu P_L b_\alpha)(\bar{s}^\beta \gamma^\mu P_L c_\beta), \\
Q_2 &= (\bar{c}^\beta \gamma_\mu P_L b_\alpha)(\bar{s}^\alpha \gamma^\mu P_L c_\beta), \\
Q_3 &= (\bar{s}^\alpha \gamma_\mu P_L b_\alpha) \sum_{q=u,d,s,c,b} \bar{q}^\beta \gamma^\mu P_L q_\beta, \\
Q_4 &= (\bar{s}^\beta \gamma_\mu P_L b_\alpha) \sum_{q=u,d,s,c,b} \bar{q}^\alpha \gamma^\mu P_L q_\beta, \\
Q_5 &= (\bar{s}^\alpha \gamma_\mu P_L b_\alpha) \sum_{q=u,d,s,c,b} \bar{q}^\beta \gamma^\mu P_R q_\beta, \\
Q_6 &= (\bar{s}^\beta \gamma_\mu P_L b_\alpha) \sum_{q=u,d,s,c,b} \bar{q}^\alpha \gamma^\mu P_R q_\beta.
\end{aligned} \tag{6.118}$$

At the subtraction point $\mu = M_W$, the coefficients are

$$C_1(M_W) = 1 + \mathcal{O}\left[\alpha_s(M_W)\right], \qquad C_{j\neq 1}(M_W) = 0 + \mathcal{O}\left[\alpha_s(M_W)\right]. \tag{6.119}$$

The operators Q_1 and Q_2 are analogous to O_1 and O_2 in the $\Delta c = 1$ nonleptonic Hamiltonian. The new operators $Q_3 - Q_6$ occur because new "penguin" diagrams shown in Fig. 6.13 occur in the renormalization of Q_1. The sum of diagrams in Fig. 6.13 is proportional to the tree-level matrix element of the operator

$$g(\bar{s}T^A \gamma_\mu P_L b) D_\nu G^{A\nu\mu}, \tag{6.120}$$

which after using the equation of motion $D_\nu G^{A\nu\mu} = g \sum_q \bar{q}\gamma^\mu T^A q$ becomes

$$g^2(\bar{s}T^A \gamma_\mu P_L b) \sum_{q=u,d,s,c,b} \bar{q}\gamma^\mu T^A q. \tag{6.121}$$

This is a linear combination of $Q_3 - Q_6$. Penguin-type diagrams with more gluons attached to the loop are finite and do not contribute to the operator renormalization.

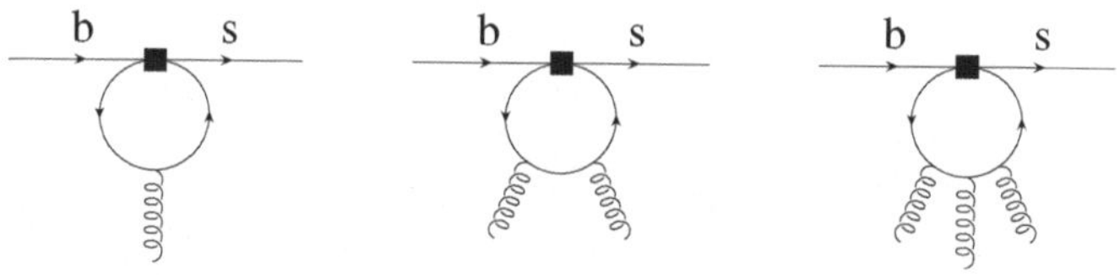

Fig. 6.13. Penguin diagrams that renormalize the weak Hamiltonian.

The coefficients of Q_{1-6} at $\mu = m_b$ are computed by using the renormalization group equation Eq. (1.134), where the anomalous dimension matrix is

$$\gamma = \frac{g^2}{8\pi^2}\begin{pmatrix} -1 & 3 & -\frac{1}{9} & \frac{1}{3} & -\frac{1}{9} & \frac{1}{3} \\ 3 & -1 & 0 & 0 & 0 & 0 \\ 0 & 0 & -\frac{11}{9} & \frac{11}{3} & -\frac{2}{9} & \frac{2}{3} \\ 0 & 0 & \frac{22}{9} & \frac{2}{3} & -\frac{5}{9} & \frac{5}{3} \\ 0 & 0 & 0 & 0 & 1 & -3 \\ 0 & 0 & -\frac{5}{9} & \frac{5}{3} & -\frac{5}{9} & -\frac{19}{3} \end{pmatrix}. \tag{6.122}$$

Solving Eq. (1.134) for the coefficients at the scale $\mu = m_b$, it is easy to see that C_1 and C_2 have the same value as in the $\Delta c = 1$ case, whereas $C_3 - C_6$ are quite small.

The operators in the OPE for the time-ordered product of weak Hamiltonians that gives $\Delta\Gamma$ must be both $\Delta s = 2$ and $\Delta b = 2$. Consequently, the lowest-dimension operators are four-quark operators, and $\Delta\Gamma$ is suppressed by $\Lambda^3_{\rm QCD}/m_b^3$ in comparison with Γ.

Neglecting the operators $Q_3 - Q_6$, we calculate the operator product for the time-ordered product in Eq. (6.116) from the imaginary part of the one-loop Feynman diagram in Fig. 6.14. This gives

$$\Delta\Gamma = \big[C_1^2\langle B_s(v)|(\bar{s}^\beta\gamma^\mu P_L b_{v\alpha})(\bar{s}^\alpha\gamma^\nu P_L b_{v\beta})|\bar{B}_s(v)\rangle + \big(3C_2^2 + 2C_1C_2\big) \\ \times \langle B_s(v)|(\bar{s}^\alpha\gamma^\mu P_L b_{v\alpha})(\bar{s}^\beta\gamma^\nu P_L b_{v\beta})|\bar{B}_s(v)\rangle\big]\,\mathrm{Im}\,\Pi_{\mu\nu}(p_b). \tag{6.123}$$

Taking the imaginary part converts the loop integration into a phase space integration for the intermediate c and $\bar{c}$ quarks:

$$\mathrm{Im}\,\Pi_{\mu\nu}(p_b) = 4G_F^2\left(V_{cb}V_{cs}^*\right)^2\int\frac{d^3p_c}{(2\pi)^3 2E_c}\frac{d^3p_{\bar{c}}}{(2\pi)^3 2E_{\bar{c}}}(2\pi)^4\,\delta^4\left(p_b - p_c - p_{\bar{c}}\right) \\ \times \mathrm{Tr}[\gamma_\mu P_L(\not{p}_c + m_c)\gamma_\nu P_L(\not{p}_{\bar{c}} - m_c)]. \tag{6.124}$$

Performing the phase space integration above yields

$$\mathrm{Im}\,\Pi_{\mu\nu}(p_b) = 4G_F^2\left(V_{cb}V_{cs}^*\right)^2 m_b^2(Ev_\mu v_\nu + Fg_{\mu\nu}), \tag{6.125}$$

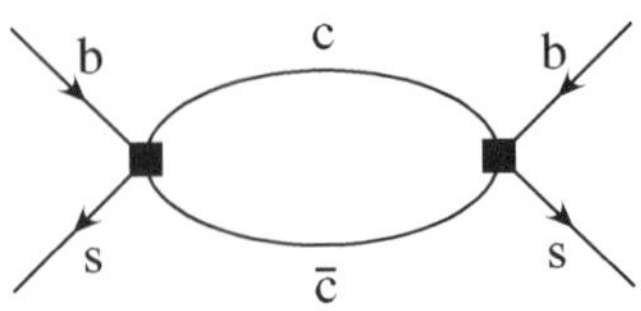

Fig. 6.14. One-loop diagram for $B_s - \bar{B}_s$ mixing.

where

$$\begin{aligned} E &= \frac{1+2\rho}{24\pi}\sqrt{1-4\rho}, \\ F &= -\frac{1-\rho}{24\pi}\sqrt{1-4\rho}, \end{aligned} \tag{6.126}$$

and $\rho = m_c^2/m_b^2$. Putting the above results together gives

$$\begin{aligned} \Delta\Gamma = \; & \frac{G_F^2\left(V_{cb}V_{cs}^*\right)^2 m_b^2}{6\pi}\sqrt{1-4\rho} \\ & \times \Big\{\big[C_1^2\langle B_s(v)|(\bar{s}^\beta P_R b_{v\alpha})\,(\bar{s}^\alpha P_R b_{v\beta})|\bar{B}_s(v)\rangle \\ & + \left(3C_2^2 + 2C_1C_2\right) \\ & \times \langle B_s(v)|(\bar{s}^\beta P_R b_{v\beta})\,(\bar{s}^\alpha P_R b_{v\alpha})|\bar{B}_s(v)\rangle\big](1+2\rho) \\ & - \left(C_1^2 + 3C_2^2 + 2C_1C_2\right) \\ & \times \langle B_s(v)|(\bar{s}^\beta \gamma^\mu P_L b_{v\beta})\,(\bar{s}^\alpha \gamma_\mu P_L b_{v\alpha})|\bar{B}_s(v)\rangle(1-\rho)\Big\}. \end{aligned} \tag{6.127}$$

One of the four-quark operators can be eliminated by using the Fierz identity:

$$\begin{aligned} & (\bar{s}^\alpha \gamma^\mu P_L b_\alpha)(\bar{s}^\beta \gamma^\nu P_L b_\beta) + (\bar{s}^\beta \gamma^\mu P_L b_\alpha)(\bar{s}^\alpha \gamma^\nu P_L b_\beta) \\ & \quad = \frac{1}{2} g^{\mu\nu}(\bar{s}^\alpha \gamma^\lambda P_L b_\alpha)(\bar{s}^\beta \gamma_\lambda P_L b_\beta). \end{aligned} \tag{6.128}$$

Making the transition to HQET and contracting with $v_\mu v_\nu$, we find this Fierz identity gives

$$\begin{aligned} & (\bar{s}^\alpha P_R b_{v\alpha})(\bar{s}^\beta P_R b_{v\beta}) + (\bar{s}^\beta P_R b_{v\alpha})(\bar{s}^\alpha P_R b_{v\beta}) \\ & \quad = \frac{1}{2}(\bar{s}^\alpha \gamma^\lambda P_L b_{v\alpha})(\bar{s}^\beta \gamma_\lambda P_L b_{v\beta}). \end{aligned} \tag{6.129}$$

Using this, Eq. (6.127) becomes

$$\begin{aligned} \Delta\Gamma = & -\frac{G_F^2\left(V_{cb}V_{cs}^*\right)^2 m_b^2}{6\pi}\sqrt{1-4\rho} \times \Big\{\left(-C_1^2 + 2C_1C_2 + 3C_2^2\right)(1+2\rho) \\ & \times \langle B_s(v)|(\bar{s}^\beta\, P_R b_{v\beta})\,(\bar{s}^\alpha\, P_R b_{v\alpha})|\bar{B}_s(v)\rangle \\ & + \left[\frac{1}{2}C_1^2(1-4\rho) + \left(3C_2^2 + 2C_1C_2\right)(1-\rho)\right] \\ & \times \langle B_s(v)|(\bar{s}^\beta \gamma^\mu\, P_L b_{v\beta})\,(\bar{s}^\alpha \gamma_\mu\, P_L b_{v\alpha})|\bar{B}_s(v)\rangle\Big\}. \end{aligned} \tag{6.130}$$

Estimates of the matrix elements in this equation suggest that $|\Delta\Gamma/\Gamma_{B_s}|$ is ~0.1.

6.10 Problems

1. At fixed q^2, show that the structure functions $F_{1,2}(\omega, q^2)$ defined in Sec. 1.8 have cuts on the real ω axis for $|\omega| \geq 1$. Also show that the discontinuity across the positive ω cut is given by Eq. (1.165).
2. Derive Eqs. (6.10) and (6.11).
3. Define the parton-level dimensionless energy and invariant mass variables $\hat{E}_0$ and $\hat{s}_0$ by

$$\hat{E}_0 = v \cdot (p_b - q)/m_b = 1 - v \cdot \hat{q},$$
$$\hat{s}_0 = (p_b - q)^2/m_b^2 = 1 - 2v \cdot \hat{q} + \hat{q}^2.$$

The hadronic energy E_H and invariant mass s_H are given by

$$E_H = v \cdot (p_B - q) = m_B - v \cdot q,$$
$$s_H = (p_B - q)^2 = m_B^2 - 2m_B v \cdot q + q^2.$$

(a) Show that E_H and s_H are related to the parton-level quantities by

$$E_H = \bar{\Lambda} - \frac{\lambda_1 + 3\lambda_2}{2m_B} + \left(m_B - \bar{\Lambda} + \frac{\lambda_1 + 3\lambda_2}{2m_B}\right)\hat{E}_0 + \cdots$$
$$s_H = m_c^2 + \bar{\Lambda}^2 + \left(m_B^2 - 2\bar{\Lambda}m_B + \bar{\Lambda}^2 + \lambda_1 + 3\lambda_2\right)(\hat{s}_0 - \rho) + (2\bar{\Lambda}m_B - 2\bar{\Lambda}^2 - \lambda_1 - 3\lambda_2)\hat{E}_0 + \cdots,$$

where the ellipses denote terms of higher order in $1/m_B$.

(b) For the $b \to u$ case, set $m_c = 0$ in the above and show that

$$\langle \hat{s}_0 \rangle = \frac{13\lambda_1}{20m_b^2} + \frac{3\lambda_2}{4m_b^2},$$
$$\langle \hat{E}_0 \rangle = \frac{13\lambda_1}{40m_b^2} + \frac{63\lambda_2}{40m_b^2},$$

where the symbol $\langle\cdot\rangle$ denotes an average over the decay phase space.

(c) Use the previous results to show that

$$\langle s_H \rangle = m_B^2\left[\frac{7\bar{\Lambda}}{10m_B} + \frac{3}{10m_B^2}(\bar{\Lambda}^2 + \lambda_1 - \lambda_2)\right].$$

4. Define

$$T_{\mu\nu} = -i \int d^4x\, e^{-iq\cdot x} \frac{\langle \bar{B}|T\left[J_\mu^\dagger(x)J_\nu(0)\right]|\bar{B}\rangle}{2m_B},$$

where J is a $b \to c$ vector or axial current. An operator product expansion of $T_{\mu\nu}$ in the zero-recoil case $\mathbf{q} = 0$ yields

$$\frac{1}{3}T_{ii}^{AA} = \frac{1}{\varepsilon} - \frac{(\lambda_1 + 3\lambda_2)(m_b - 3m_c)}{6m_b^2\varepsilon\,(2m_c + \varepsilon)} + \frac{4\lambda_2 m_b - (\lambda_1 + 3\lambda_2)(m_b - m_c - \varepsilon)}{m_b\varepsilon^2\,(2m_c + \varepsilon)},$$
$$\frac{1}{3}T_{ii}^{VV} = \frac{1}{2m_c + \varepsilon} - \frac{(\lambda_1 + 3\lambda_2)(m_b + 3m_c)}{6m_b^2\varepsilon\,(2m_c + \varepsilon)} + \frac{4\lambda_2 m_b - (\lambda_1 + 3\lambda_2)(m_b - m_c - \varepsilon)}{m_b\varepsilon\,(2m_c + \varepsilon)^2},$$

where $\varepsilon = m_b - m_c - q_0$.

(a) Use these results to deduce the sum rules

$$\frac{1}{6m_B}\sum_X (2\pi)^3\,\delta^3(\mathbf{p}_X)\,|\langle X|A_i|\bar{B}\rangle|^2 = 1 - \frac{\lambda_2}{m_c^2} + \frac{\lambda_1+3\lambda_2}{4}\left(\frac{1}{m_c^2}+\frac{1}{m_b^2}+\frac{2}{3m_bm_c}\right),$$

$$\frac{1}{6m_B}\sum_X (2\pi)^3\,\delta^3(\mathbf{p}_X)\,|\langle X|V_i|\bar{B}\rangle|^2 = \frac{\lambda_2}{m_c^2} - \frac{\lambda_1+3\lambda_2}{4}\left(\frac{1}{m_c^2}+\frac{1}{m_b^2}-\frac{2}{3m_bm_c}\right).$$

(b) Use part (a) to deduce the bounds

$$h_{A_1}^2(1) \le 1 - \frac{\lambda_2}{m_c^2} + \frac{\lambda_1+3\lambda_2}{4}\left(\frac{1}{m_c^2}+\frac{1}{m_b^2}+\frac{2}{3m_bm_c}\right),$$

$$0 \le \frac{\lambda_2}{m_c^2} - \frac{\lambda_1+3\lambda_2}{4}\left(\frac{1}{m_c^2}+\frac{1}{m_b^2}-\frac{2}{3m_bm_c}\right).$$

5. Use the results of Sec. 6.2 to derive the double differential decay rate in Eq. (6.57).
6. Calculate the renormalization of $Q_1 - Q_6$ and verify the anomalous dimension matrix in Eq. (6.122).
7. Suppose the effective Hamiltonian for semileptonic weak B decay is

$$H_W = \frac{G_F}{\sqrt{2}} V_{cb}(\bar{c}\gamma_\mu b)(\bar{e}\gamma^\mu \nu_e).$$

Perform an OPE on the time-ordered product of vector currents and deduce the nonperturbative $1/m_b^2$ corrections to $d\Gamma/d\hat{q}^2\,dy$.

6.11 References

Inclusive semileptonic $\bar{B}$ decays were studied in:

J. Chay, H. Georgi, and B. Grinstein, Phys. Lett. B247 (1990) 399
I.I. Bigi, M.A. Shifman, N.G. Uraltsev, and A.I. Vainshtein, Phys. Rev. Lett. 71 (1993) 496
A.V. Manohar and M.B. Wise, Phys. Rev. D49 (1994) 1310
B. Blok, L. Koyrakh, M.A. Shifman, and A.I. Vainshtein, Phys. Rev. D49 (1994) 3356 [erratum: ibid D50 (1994) 3572]
T. Mannel, Nucl. Phys. B413 (1994) 396

Inclusive semileptonic $\bar{B} \to \tau$ decay was studied in:

A.F. Falk, Z. Ligeti, M. Neubert, and Y. Nir, Phys. Lett. B326 (1994) 145
L. Koyrakh, Phys. Rev. D49 (1994) 3379
S. Balk, J.G. Korner, D. Pirjol, and K. Schilcher, Z. Phys. C64 (1994) 37

Perturbative QCD corrections to the $\bar{B}$ semileptonic decay rate were computed in:

Q. Hokim and X.Y. Pham, Ann. Phys. 155 (1984) 202, Phys. Lett. B122 (1983) 297
M. Jezabek and J.H. Kuhn, Nucl. Phys. B320 (1989) 20
A. Czarnecki and M. Jezabek, Nucl. Phys. B427 (1994) 3
M.E. Luke, M.J. Savage, and M.B. Wise, Phys. Lett. B343 (1995) 329, B345 (1995) 301
A. Czarnecki and K. Melnikov, Phys. Rev. Lett. 78 (1997) 3630, hep-ph/9804215
M. Gremm and I. Stewart, Phys. Rev. D55 (1997) 1226

For $1/m_b^3$ corrections to the $\bar{B}$ semileptonic decay rate, see:

M. Gremm and A. Kapustin, Phys. Rev. D55 (1997) 6924

The shape function in the endpoint region was studied in:

M. Neubert, Phys. Rev. D49 (1994) 4623
I.I. Bigi, M.A. Shifman, N.G. Uraltsev, and A.I. Vainshtein, Int. J. Mod. Phys. A9 (1994) 2467

For discussions on quark–hadron duality, see, for example:

E.C. Poggio, H.R. Quinn, and S. Weinberg, Phys. Rev. D13 (1976) 1958
C.G. Boyd, B. Grinstein, and A.V. Manohar, Phys. Rev. D54 (1996) 2081
B. Blok, M.A. Shifman, and D.-X. Zhang, Phys. Rev. D57 (1998) 2691
I.I. Bigi, M.A. Shifman, N.G. Uraltsev, and A.I. Vainshtein, hep-ph/9805241
N. Isgur, hep-ph/9809279

Sum rules were discussed in:

J.D. Bjorken, Nucl. Phys. B371 (1992) 111
N. Isgur and M.B. Wise, Phys. Rev. D43 (1991) 819
M.B. Voloshin, Phys. Rev. D46 (1992) 3062
I.I. Bigi, M.A. Shifman, N.G. Uraltsev, and A.I. Vainshtein, Phys. Rev. D52 (1995) 196
A. Kapustin, Z. Ligeti, M.B. Wise, and B. Grinstein, Phys. Lett. B375 (1996) 327
I.I. Bigi, M.A. Shifman, and N.G. Uraltsev, Ann. Rev. Nucl. Part. Sci. 47 (1997) 591
C.G. Boyd and I.Z. Rothstein, Phys. Lett. B395 (1997) 96, B420 (1998) 350
C.G. Boyd, Z. Ligeti, I.Z. Rothstein, and M.B. Wise, Phys. Rev. D55 (1997) 3027
A. Czarnecki, K. Melnikov, and N.G. Uraltsev, Phys. Rev. D57 (1998) 1769

Inclusive nonleptonic decays were studied in:

M.A. Shifman and M.B. Voloshin, Sov. J. Nucl. Phys. 41 (1985) 120 [Yad. Fiz. 41 (1985) 187]
I.I. Bigi, N.G. Uraltsev, and A.I. Vainshtein, Phys. Lett. B293 (1992) 430 [erratum: ibid B297 (1993) 477]
B. Blok and M.A. Shifman, Nucl. Phys. B399 (1993) 441, B399 (1993) 4591
M. Neubert and C.T. Sachrajda, Nucl. Phys. B483 (1997) 339

For calculations of $B_s - \bar{B}_s$ mixing, see:

M.A. Shifman and M.B. Voloshin, Sov. J. Nucl. Phys. 45 (1987) 292
M.B. Voloshin, N.G. Uraltsev, V.A. Khoze, and M.A. Shifman, Sov. J. Nucl. Phys. 46 (1987) 382
M. Beneke, G. Buchalla, and I. Dunietz, Phys. Rev. D54 (1996) 4419

For applications to inclusive rare decays, see:

A.F. Falk, M.E. Luke, and M.J. Savage, Phys. Rev. D49 (1994) 3367
A. Ali, G. Hiller, L.T. Handoko, and T. Morozumi, Phys. Rev. D55 (1997) 4105

The hadronic mass spectrum was considered in:

A.F. Falk, M.E. Luke, and M.J. Savage, Phys. Rev. D53 (1996) 2491, D53 (1996) 6316
R.D. Dikeman and N.G. Uraltsev, Nucl. Phys. B509 (1998) 378
A.F. Falk, Z. Ligeti, and M.B. Wise, Phys. Lett. B406 (1997) 225

Index

9 789390 013968

Printed at Repro India Ltd.